Lecture Notes in Computer Science　　9525

Commenced Publication in 1973
Founding and Former Series Editors:
Gerhard Goos, Juris Hartmanis, and Jan van Leeuwen

More information about this series at http://www.springer.com/series/7407

Aske Plaat · Jaap van den Herik
Walter Kosters (Eds.)

Advances in Computer Games

14th International Conference, ACG 2015
Leiden, The Netherlands, July 1–3, 2015
Revised Selected Papers

 Springer

Editors
Aske Plaat
Leiden University
Leiden, Zuid-Holland
The Netherlands

Jaap van den Herik
Leiden University
Leiden, Zuid-Holland
The Netherlands

Walter Kosters
Leiden University
Leiden, Zuid-Holland
The Netherlands

ISSN 0302-9743   ISSN 1611-3349   (electronic)
Lecture Notes in Computer Science
ISBN 978-3-319-27991-6   ISBN 978-3-319-27992-3   (eBook)
DOI 10.1007/978-3-319-27992-3

Library of Congress Control Number: 2015958336

LNCS Sublibrary: SL1 – Theoretical Computer Science and General Issues

Printed on acid-free paper

This Springer imprint is published by SpringerNature
The registered company is Springer International Publishing AG Switzerland

Preface

This book contains the papers of the 14th Advances in Computer Games Conference (ACG 2015) held in Leiden, The Netherlands. The conference took place during July 1–3, 2015, in conjunction with the 18th Computer Olympiad and the 21st World Computer-Chess Championship.

The Advances in Computer Games conference series is a major international forum for researchers and developers interested in all aspects of artificial intelligence and computer game playing. As with many of the recent conferences, this conference saw a large number of papers with progress in Monte Carlo tree search. However, the Leiden conference also showed diversity in the topics of papers, e.g., by including papers on thorough investigations on search, theory, complexity, games (performance and pre-filing), and machine learning. Earlier conferences took place in London (1975), Edinburgh (1978), London (1981, 1984), Noordwijkerhout (1987), London (1990), Maastricht (1993, 1996), Paderborn (1999), Graz (2003), Taipei (2005), Pamplona (2009), and Tilburg (2011).

The Program Committee (PC) was pleased to see that so much progress was made in new games and that new techniques were added to the recorded achievements. In this conference, 36 authors submitted a paper. Each paper was sent to at least three reviewers. If conflicting views on a paper were reported, the reviewers themselves arrived at a final decision. With the help of external reviewers (see after the preface), the PC accepted 22 papers for presentation at the conference and publication in these proceedings. As usual we informed the authors that they submitted their contribution to a post-conference editing process. The two-step process is meant (a) to give authors the opportunity to include the results of the fruitful discussion after the lecture in their paper, and (b) to maintain the high-quality threshold of the ACG series. The authors enjoyed this procedure.

The aforementioned set of 22 papers covers a wide range of computer games and many different research topics. We grouped the topics into the following five classes according to the order of publication: Monte Carlo tree search (MCTS) and its enhancements (six papers), theoretical aspects and complexity (five papers), analysis of game characteristics (five papers), search algorithms (three papers), and machine learning (three papers).

We hope that the readers will enjoy the research efforts presented by the authors. Here, we reproduce brief characterizations of the 22 contributions largely relying on the text as submitted by the authors. The idea is to show a connection between the contributions and insights into the research progress.

Monte Carlo Tree Search

The authors of the first publication "Adaptive Playouts in Monte Carlo Tree Search with Policy Gradient Reinforcement Learning" received the ACG 2015 Best Paper Award. The paper is written by Tobias Graf and Marco Platzner. As the title suggests, the paper introduces an improvement in the playout phase of MCTS. If the playout policy evaluates a position wrongly then cases may occur where the tree search faces severe difficulties to find the correct move; a large search space may be entered unintentionally. The paper explores adaptive playout policies. The intention is to improve the playout policy *during* the tree search. With the help of policy-gradient reinforcement learning techniques the program optimizes the playout policy to arrive at better evaluations. The authors provide tests for computer Go and show an increase in playing strength of more than 100 ELO points.

"Early Playout Termination in MCTS," authored by Richard Lorentz, also deals with the playout phase of MCTS. Nowadays minimax-based and MCTS-based searches are seen as competitive and incompatible approaches. For example, it is generally agreed that chess and checkers require a minimax approach while Go and Havannah are better off by MCTS. However, a hybrid technique is also possible. It works by stopping the random MCTS playouts early and using an evaluation function to determine the winner of the playout. This algorithm is called MCTS-EPT (MCTS with early playout termination), and is studied in this paper by using MCTS-EPT programs written for Amazons, Havannah, and Breakthrough.

"Playout Policy Adaptation for Games," a contribution by Tristan Cazenave, is the third paper on playouts in MCTS. For general game playing, MCTS is the algorithm of choice. The paper proposes to learn a playout policy online in order to improve MCTS for general game playing. The resulting algorithm is named Playout Policy Adaptation, and was tested on Atarigo, Breakthrough, Misere Breakthrough, Domineering, Misere Domineering, Go, Knightthrough, Misere Knightthrough, Nogo, and Misere Nogo. For most of these games, Playout Policy Adaptation is better than UCT for MCTS with a uniform random playout policy, with the notable exceptions of Go and Nogo.

"Strength Improvement and Analysis for an MCTS-Based Chinese Dark Chess Program" is authored by Chu-Hsuan Hsueh, I-Chen Wu, Wen-Jie Tseng, Shi-Jim Yen, and Jr-Chang Chen. The paper describes MCTS and its application in the game of Chinese Dark Chess (CDC), with emphasis on playouts. The authors study how to improve and analyze the playing strength of an MCTS-based CDC program, named DARKKNIGHT, which won the CDC tournament in the 17th Computer Olympiad. They incorporated three recent techniques, early playout terminations, implicit minimax backups, and quality-based rewards. For early playout terminations, playouts end when reaching states with likely outcomes. Implicit minimax backups use heuristic evaluations to help guide selections of MCTS. Quality-based rewards adjust rewards based on online collected information. Experiments show that the win rates against the original DARKKNIGHT are 60 %, 70 %, and 59 %, respectively, when incorporating the three techniques. By incorporating all three together, the authors obtain a win rate of 77 %.

"LinUCB Applied to Monte Carlo Tree Search" is written by Yusaku Mandai and Tomoyuki Kaneko. This paper deals with the selection function of MCTS; as is well known, UCT is a standard selection method for MCTS algorithms. The study proposes a family of LinUCT algorithms (linear UCT algorithm) that incorporate LinUCB into MCTS algorithms. LinUCB is a recently developed method. It generalizes past episodes by ridge regression with feature vectors and rewards. LinUCB outperforms UCB1 in contextual multi-armed bandit problems. The authors introduce a straightforward application of LinUCB, called $LinUCT_{PLAIN}$. They show that it does not work well owing to the minimax structure of game trees. Subsequently, they present $LinUCT_{RAVE}$ and $LinUCT_{FP}$ by further incorporating two existing techniques, rapid action value estimation (RAVE) and feature propagation (FP). Experiments were conducted by a synthetic model. The experimental results indicate that $LinUCT_{RAVE}$, $LinUCT_{FP}$, and their combination $LinUCT_{RAVE-FP}$ all outperform UCT, especially when the branching factor is relatively large.

"Adapting Improved Upper Confidence Bounds for Monte-Carlo Tree Search" is authored by Yun-Ching Liu and Yoshimasa Tsuruoka. The paper asserts that the UCT algorithm, which is based on the UCB algorithm, is currently the most widely used variant of MCTS. Recently, a number of investigations into applying other bandit algorithms to MCTS have produced interesting results. In their paper, the authors investigate the possibility of combining the improved UCB algorithm, proposed by Auer et al., with MCTS. The paper describes the Mi-UCT algorithm, which applies the modified UCB algorithm to trees. The performance of Mi-UCT is demonstrated on the games of 9×9 Go and 9×9 NoGo. It was shown to outperform the plain UCT algorithm when only a small number of playouts are given, while performing roughly on the same level when more playouts are available.

Theory and Complexity

"On Some Random Walk Games with Diffusion Control" is written by Ingo Althöfer, Matthias Beckmann, and Friedrich Salzer. This paper is not directly about MCTS, but contains a theoretical study of an element related to MCTS: the effect of random movements. The paper remarks that random walks with discrete time steps and discrete state spaces have widely been studied for several decades. The authors investigate such walks as games with "diffusion control": a player (controller) with certain intentions influences the random movements of a particle. In their models the controller decides only about the step size for a single particle. It turns out that this small amount of control is sufficient to cause the particle to stay in "premium regions" of the state space with surprisingly high probabilities.

"On Some Evacuation Games with Random Walks" is a contribution by Matthias Beckmann. The paper contains a theoretical study of a game with randomness. A single player game is considered; a particle on a board has to be steered toward evacuation cells. The actor has no direct control over this particle but may indirectly influence the movement of the particle by blockades. Optimal blocking strategies and the recurrence property are examined experimentally. It is concluded that the random walk of the game is recurrent. Furthermore, the author describes the average time in which an evacuation cell is reached.

"Go Complexities" is authored by Abdallah Saffidine, Olivier Teytaud, and Shi-Jim Yen. The paper presents a theoretical study about the complexity of Go. The game of Go is often said to be EXPTIME-complete. This result refers to classic Go under Japanese rules, but many variants of the problem exist and affect the complexity. The authors survey what is known about the computational complexity of Go and highlight challenging open problems. They also propose new results. In particular, the authors show that Atari-Go is PSPACE-complete and that hardness results for classic Go carry over to their partially observable variant.

"Crystallization of Domineering Snowflakes" is written by Jos W.H.M. Uiterwijk. The paper contains a theoretical analysis of the game of Domineering. The author presents a combinatorial game-theoretic analysis of special Domineering positions. He investigates complex positions that are aggregates of simpler components and linked via bridging squares. Two theorems are introduced. They state that an aggregate of two components has as its game-theoretic value the sum of the values of the components. These theorems are then extended to deal with the case of multiple-connected networks of components. As an application, an interesting Domineering position is introduced, consisting of a $*2$ subgame (star-two subgame) with four $*$ components attached: the so-called Snowflake position. The author then shows how from this Snowflake, larger chains of Snowflakes can be built with known values, including flat networks of Snowflakes (a kind of crystallization).

"First Player's Cannot-Lose Strategies for Cylinder-Infinite-Connect-Four with Widths 2 and 6" is a contribution by Yoshiaki Yamaguchi and Todd Neller. The paper introduces a theoretical result on a variant of Connect-Four. Cylinder-Infinite-Connect-Four is Connect-Four played on a cylindrical square grid board with infinite row height and columns that cycle about its width. In previous work by the same authors, the first player's cannot-lose strategies were discovered for all widths except 2 and 6, and the second player's cannot-lose strategies have been discovered for all widths except 6 and 11. In this paper, the authors show the first player's cannot-lose strategies for widths 2 and 6.

Analysis of Game Characteristics

"Development of a Program for Playing Progressive Chess" is written by Vito Janko and Matej Guid. They present the design of a computer program for playing Progressive Chess. In this game, rather than just making one move per turn, players play progressively longer series of moves. The program follows the generally recommended strategy for this game, which consists of three phases: (1) looking for possibilities to checkmate the opponent, (2) playing generally good moves when no checkmate can be found, and (3) preventing checkmates from the opponent. In their paper, the authors focus on efficiently searching for checkmates, putting to test various heuristics for guiding the search. They also present the findings of self-play experiments between different versions of the program.

"A Comparative Review of Skill Assessment: Performance, Prediction, and Profiling" is a contribution by Guy Haworth, Tamal Biswas, and Ken Regan. The paper presents results on the skill assessment of chess players. The authors argue that the assessment of chess players is both an increasingly attractive opportunity and an unfortunate necessity. Skill assessment is important for the chess community to limit potential reputational damage by inhibiting cheating and unjustified accusations of cheating. The paper argues that there has been a recent rise in both. A number of counter-intuitive discoveries have been made by benchmarking the intrinsic merit of players' moves. The intrinsic merits call for further investigation. Is Capablanca actually, objectively the most accurate world champion? Has ELO-rating inflation not taken place? Stimulated by FIDE/ACP, the authors revisit the fundamentals of the subject. They aim at a framework suitable for improved standards of computational experiments and more precise results. The research is exemplary for other games and domains. They look at chess as a demonstrator of good practice, including (1) the rating of professionals who make high-value decisions under pressure, (2) personnel evaluation by multichoice assessment, and (3) the organization of crowd-sourcing in citizen science projects. The "3P" themes of performance, prediction, and profiling pervade all these domains.

"Boundary Matching for Interactive Sprouts" is written by Cameron Browne. The paper introduces results for the game of Sprouts. The author states that the simplicity of the pen-and-paper game Sprouts hides a surprising combinatorial complexity. He then describes an optimization called boundary matching that accommodates this complexity. It allows move generation for Sprouts games of arbitrary size at interactive speeds. The representation of Sprouts positions plays an important role. "Draws, Zugzwangs, and PSPACE-Completeness in the Slither Connection Game" is written by Édouard Bonnet, Florian Jamain, and Abdallah Saffidine. The paper deals with a connection game: Slither. Two features set Slither apart from other connection games: (1) previously played stones can be displaced and (2) some stone configurations are forbidden. The standard goal is still connecting opposite edges of a board. The authors show that the interplay of the peculiar mechanics results in a game with a few properties unexpected among connection games; for instance, the existence of mutual Zugzwangs. The authors establish that (1) there are positions where one player has no legal move, and (2) there is no position where both players lack a legal move. The latter implies that the game cannot end in a draw. From the viewpoint of computational complexity, it is shown that the game is PSPACE-complete. The displacement rule can indeed be tamed so as to simulate a Hex game on a Slither board.

"Constructing Pin Endgame Databases for the Backgammon Variant Plakoto" is authored by Nikolaos Papahristou and Ioannis Refanidis. The paper deals with the game of Plakoto, a variant of backgammon that is popular in Greece. It describes the ongoing project PALAMEDES, which builds expert bots that can play backgammon variants. So far, the position evaluation relied only on self-trained neural networks. The authors report their first attempt to augment PALAMEDES with databases for certain endgame positions for the backgammon variant Plakoto. They mention five databases containing 12,480,720 records in total that can calculate accurately the best move for roughly $3.4 \cdot 10^{15}$ positions.

Search Algorithms

"Reducing the Seesaw Effect with Deep Proof-Number Search" is a contribution by Taichi Ishitobi, Aske Plaat, Hiroyuki Iida, and Jaap van den Herik. The paper studies the search algorithm Proof-Number (PN) search. In the paper, DeepPN is introduced. DeepPN is a modified version of PN search, providing a procedure to handle the seesaw effect. DeepPN employs two values associated with each node: the usual proof number and a deep value. The deep value of a node is defined as the depth to which each child node has been searched. By mixing the proof numbers and the deep value, DeepPN works with two characteristics: the best-first manner of search (equal to the original PN search) and the depth-first manner. By adjusting a parameter (called R in this paper) one can choose between best-first or depth-first behavior. In their experiments, the authors try to find a balance between both manners of searching. As it turns out, best results are obtained at an R value in between the two extremes of best-first search (original PN search) and depth-first search.

"Feature Strength and Parallelization of Sibling Conspiracy Number Search" is written by Jakub Pawlewicz and Ryan Hayward. The paper studies a variant of Conspiracy Number Search, a predecessors of Proof-Number search, with different goals. Recently the authors introduced Sibling Conspiracy Number Search (SCNS). This algorithm is not based on evaluation of the leaf nodes of the search tree but, it handles for each node, the relative evaluation scores of all children of that node. The authors implemented an SCNS Hex bot. It showed the strength of SCNS features: most critical is the initialization of the leaves via a multi-step process. The authors also investigated a simple parallel version of SCNS: it scales well for two threads but was less efficient for four or eight threads.

"Parameter-Free Tree Style Pipeline in Asynchronous Parallel Game-Tree Search" is authored by Shu Yokoyama, Tomoyuki Kaneko, and Tetsuro Tanaka. This paper describes results in parameter-free parallel search algorithms. The paper states that asynchronous parallel game-tree search methods are effective in improving playing strength by using many computers connected through relatively slow networks. In game-position parallelization, a master program manages the game tree and distributes positions in the tree to workers. Then, each worker asynchronously searches for the best move and evaluation in the assigned position. The authors present a new method for constructing an appropriate master tree that provides more important moves with more workers on their subtrees to improve playing strength. Their contribution goes along with two advantages: (1) it is parameter free in that users do not need to tune parameters through trial and error, and (2) the efficiency is suitable even for short-time matches, such as one second per move. The method was implemented in a top-level chess program (STOCKFISH). Playing strength was evaluated through self-plays. The results show that playing strength improves up to 60 workers.

Machine Learning

"Transfer Learning by Inductive Logic Programming" is a contribution by Yuichiro Sato, Hiroyuki Iida, and Jaap van den Herik. The paper studies Transfer Learning between different types of games. In the paper, the authors propose a Transfer Learning method by Inductive Logic Programming for games. The method generates general knowledge from a game, and specifies the knowledge in the form of heuristic functions, so that it is applicable in another game. This is the essence of Transfer Learning. The authors illustrate the working of Transfer Learning by taking knowledge from Tic-tac-toe and transferring it to Connect-Four and Connect-Five.

"Developing Computer Hex Using Global and Local Evaluation Based on Board Network Characteristics" is written by Kei Takada, Masaya Honjo, Hiroyuki Iizuka and Masahito Yamamoto. The paper describes a learning method for evaluation functions. The authors remark that one of the main approaches to develop a computer Hex program (in the early days) was to use an evaluation function of the electric circuit model. However, such a function evaluates the board states from one perspective only. Moreover this method has recently been defeated by MCTS approaches. In the paper, the authors propose a novel evaluation function that uses network characteristics to capture features of the board states from *two* perspectives. The proposed evaluation function separately evaluates (1) the board network and (2) the shortest path network using betweenness centrality. Then the results of these evaluations are combined. Furthermore, the proposed method involves changing the ratio between global and local evaluations through a Support Vector Machine (SVM). The new method yields an improved strategy for Hex. The resultant program, called Ezo, is tested against the world-champion Hex program called MoHex, and the results show that the method is superior to the 2011 version of MoHex on an 11×11 board.

"Machine-Learning of Shape Names for the Game of Go" is authored by Kokolo Ikeda, Takanari Shishido, and Simon Viennot. The paper discusses the learning of shape names in Go. It starts stating that Computer Go programs with only a 4-stone handicap have recently defeated professional humans. The authors argue that by this perfor-mance the strength of Go programs is sufficiently close to that of humans, so that a new target in artificial intelligence is at stake, namely, developing programs able to provide commentary on Go games. A fundamental difficulty in achieving the new target is learning the terminology of Go, which is often not well defined. An example is the problem of naming shapes such as Atari, Attachment, or Hane. In their research, the goal is to allow a program to label relevant moves with an associated shape name. The authors use machine learning to deduce these names based on local patterns of stones. First, strong amateur players recorded for each game move the associated shape name, using a pre-selected list of 71 terms. Next, these records were used to train a supervised machine learning algorithm. The result was a program able to output the shape name from the local patterns of stones. Humans agree on a shape name with a performance rate of about 82 %. The algorithm achieved a similar performance, picking the name most preferred by humans with a rate of also about 82 %. The authors state that this is a first step toward a program that is able to communicate with human players in a game review or match.

This book would not have been produced without the help of many persons. In particular, we would like to mention the authors and the reviewers for their help. Moreover, the organizers of the three events in Leiden (see the beginning of this preface) have contributed substantially by bringing the researchers together. Without much emphasis, we recognize the work by the committees of the ACG 2015 as essential for this publication. One exception is made for Joke Hellemons, who is gratefully thanked for all services to our games community. Finally, the editors happily recognize the generous sponsors AEGON, NWO, the Museum Boerhaave, SurfSARA, the Municipality of Leiden, the Leiden Institute of Advanced Computer Science, the Leiden Centre of Data Science, ICGA, and Digital Games Technology.

September 2015

Aske Plaat
Jaap van den Herik
Walter Kosters

Organization

Executive Committee

Editors

Aske Plaat
Jaap van den Herik
Walter Kosters

Program Co-chairs

Aske Plaat
Jaap van den Herik
Walter Kosters

Organizing Committee

Johanna Hellemons
Mirthe van Gaalen
Aske Plaat
Jaap van den Herik
Eefje Kruis Voorberge
Marloes van der Nat
Jan van Rijn
Jonathan Vis

List of Sponsors

AEGON
NWO (Netherlands Organization of Scientific Research)
Museum Boerhaave
SurfSARA
Leiden Institute of Advanced Computer Science
Leiden Centre of Data Science
Leiden Faculty of Science
ISSC
ICGA
Digital Games Technology
The Municipality of Leiden

Program Committee

Ingo Althöfer
Petr Baudis
Yngvi Björnsson
Bruno Bouzy
Ivan Bratko
Cameron Browne
Tristan Cazenave
Bo-Nian Chen
Jr-Chang Chen
Paolo Ciancarini
David Fotland
Johannes Fürnkranz
Michael Greenspan
Reijer Grimbergen
Matej Guid
Dap Hartmann
Tsuyoshi Hashimoto
Guy Haworth
Ryan Hayward

Jaap van den Herik
Hendrik Jan Hoogeboom
Shun-chin Hsu
Tsan-Sheng Hsu
Hiroyuki Iida
Tomoyuki Kaneko
Graham Kendall
Akihiro Kishimoto
Walter Kosters
Yoshiyuki Kotani
Richard Lorentz
Ulf Lorenz
John-Jules Meyer
Martin Müller
Todd Neller
Pim Nijssen
Jakub Pawlewicz
Aske Plaat
Christian Posthoff

Ben Ruijl
Alexander Sadikov
Jahn Saito
Maarten Schadd
Richard Segal
David Silver
Pieter Spronck
Nathan Sturtevant
Olivier Teytaud
Yoshimasa Tsuruoka
Jos Uiterwijk
Peter van Emde Boas
Jan van Zanten
Jonathan Vis
Mark Winands
Thomas Wolf
I-Chen Wu
Shi-Jim Yen

The Advances in Computer Chess/Games Books

The series of Advances in Computer Chess (ACC) Conferences started in 1975 as a complement to the World Computer-Chess Championships, for the first time held in Stockholm in 1974. In 1999, the title of the conference changed from ACC to ACG (Advances in Computer Games). Since 1975, 14 ACC/ACG conferences have been held. Below we list the conference places and dates together with the publication; the Springer publication is supplied with an LNCS series number.

London, United Kingdom (1975, March)
Proceedings of the 1st Advances in Computer Chess Conference (ACC1)
Ed. M.R.B. Clarke
Edinburgh University Press, 118 pages.

Edinburgh, United Kingdom (1978, April)
Proceedings of the 2nd Advances in Computer Chess Conference (ACC2)
Ed. M.R.B. Clarke
Edinburgh University Press, 142 pages.

London, United Kingdom (1981, April)
Proceedings of the 3rd Advances in Computer Chess Conference (ACC3)
Ed. M.R.B. Clarke
Pergamon Press, Oxford, UK, 182 pages.

London, United Kingdom (1984, April)
Proceedings of the 4th Advances in Computer Chess Conference (ACC4)
Ed. D.F. Beal
Pergamon Press, Oxford, UK, 197 pages.

Noordwijkerhout, The Netherlands (1987, April)
Proceedings of the 5th Advances in Computer Chess Conference (ACC5)
Ed. D.F. Beal
North Holland Publishing Comp., Amsterdam, The Netherlands, 321 pages.

London, United Kingdom (1990, August)
Proceedings of the 6th Advances in Computer Chess Conference (ACC6)
Ed. D.F. Beal
Ellis Horwood, London, UK, 191 pages.

Maastricht, The Netherlands (1993, July)
Proceedings of the 7th Advances in Computer Chess Conference (ACC7)
Eds. H.J. van den Herik, I.S. Herschberg, and J.W.H.M. Uiterwijk
Drukkerij Van Spijk B.V., Venlo, The Netherlands, 316 pages.

Maastricht, The Netherlands (1996, June)
Proceedings of the 8th Advances in Computer Chess Conference (ACC8)
Eds. H.J. van den Herik and J.W.H.M. Uiterwijk
Drukkerij Van Spijk B.V., Venlo, The Netherlands, 332 pages.

Paderborn, Germany (1999, June)
Proceedings of the 9th Advances in Computer Games Conference (ACG9)
Eds. H.J. van den Herik and B. Monien
Van Spijk Grafisch Bedrijf, Venlo, The Netherlands, 347 pages.

Graz, Austria (2003, November)
Proceedings of the 10th Advances in Computer Games Conference (ACG10)
Eds. H.J. van den Herik, H. Iida, and E.A. Heinz
Kluwer Academic Publishers, Boston/Dordrecht/London, 382 pages.

Taipei, Taiwan (2005, September)
Proceedings of the 11th Advances in Computer Games Conference (ACG11)
Eds. H.J. van den Herik, S-C. Hsu, T-S. Hsu, and H.H.L.M. Donkers
Springer Verlag, Berlin/Heidelberg, LNCS 4250, 372 pages.

Pamplona, Spain (2009, May)
Proceedings of the 12th Advances in Computer Games Conference (ACG12)
Eds. H.J. van den Herik and P. Spronck
Springer Verlag, Berlin/Heidelberg, LNCS 6048, 231 pages.

Tilburg, The Netherlands (2011, November)
Proceedings of the 13th Advances in Computer Games Conference (ACG13)
Eds. H.J. van den Herik and A. Plaat
Springer Verlag, Berlin/Heidelberg, LNCS 7168, 356 pages.

Leiden, The Netherlands (2015, July)
Proceedings of the 14th Advances in Computer Games Conference (ACG14)
Eds. A. Plaat, H.J. van den Herik, and W. Kosters
Springer, Heidelberg, LNCS 9525, 260 pages.

The Computers and Games Books

The series of Computers and Games (CG) Conferences started in 1998 as a complement to the well-known series of conferences in Advances in Computer Chess (ACC). Since 1998, eight CG conferences have been held. Below we list the conference places and dates together with the Springer publication (LNCS series no.)

Tsukuba, Japan (1998, November)
Proceedings of the First Computers and Games Conference (CG98)
Eds. H.J. van den Herik and H. Iida
Springer Verlag, Berlin/Heidelberg, LNCS 1558, 335 pages.

Hamamatsu, Japan (2000, October)
Proceedings of the Second Computers and Games Conference (CG2000)
Eds. T.A. Marsland and I. Frank
Springer Verlag, Berlin/Heidelberg, LNCS 2063, 442 pages.

Edmonton, Canada (2002, July)
Proceedings of the Third Computers and Games Conference (CG2002)
Eds. J. Schaeffer, M. Müller, and Y. Björnsson
Springer Verlag, Berlin/Heidelberg, LNCS 2883, 431 pages.

Ramat-Gan, Israel (2004, July)
Proceedings of the 4th Computers and Games Conference (CG2004)
Eds. H.J. van den Herik, Y. Björnsson, and N.S. Netanyahu
Springer Verlag, Berlin/Heidelberg, LNCS 3846, 404 pages.

Turin, Italy (2006, May)
Proceedings of the 5th Computers and Games Conference (CG2006)
Eds. H.J. van den Herik, P. Ciancarini, and H.H.L.M. Donkers
Springer Verlag, Berlin/Heidelberg, LNCS 4630, 283 pages.

Beijing, China (2008, September)
Proceedings of the 6th Computers and Games Conference (CG2008)
Eds. H.J. van den Herik, X. Xu, Z. Ma, and M.H.M. Winands
Springer Verlag, Berlin/Heidelberg, LNCS 5131, 275 pages.

Kanazawa, Japan (2010, September)
Proceedings of the 7th Computers and Games Conference (CG2010)
Eds. H.J. van den Herik, H. Iida, and A. Plaat
Springer Verlag, Berlin/Heidelberg, LNCS 6515, 275 pages.

Yokohama, Japan (2013, August)
Proceedings of the 8th Computers and Games Conference (CG2013)
Eds. H.J. van den Herik, H. Iida, and A. Plaat
Springer, Heidelberg, LNCS 8427, 260 pages.

Contents

Adaptive Playouts in Monte-Carlo Tree Search with Policy-Gradient Reinforcement Learning

Tobias Graf[(✉)] and Marco Platzner

University of Paderborn, Paderborn, Germany
tobiasg@mail.upb.de, platzner@upb.de

Abstract. Monte-Carlo Tree Search evaluates positions with the help of a playout policy. If the playout policy evaluates a position wrong then there are cases where the tree-search has difficulties to find the correct move due to the large search-space. This paper explores adaptive playout-policies which improve the playout-policy during a tree-search. With the help of policy-gradient reinforcement learning techniques we optimize the playout-policy to give better evaluations. We tested the algorithm in Computer Go and measured an increase in playing strength of more than 100 ELO. The resulting program was able to deal with difficult test-cases which are known to pose a problem for Monte-Carlo-Tree-Search.

1   Introduction

Monte-Carlo Tree Search (MCTS) [6] evaluates positions in a search tree by averaging the results of several random playouts. The playouts follow a fixed policy to choose moves until they reach a terminal position where the result follows from the rules of the game. The expected outcome of a playout therefore determines the quality of evaluation used in the tree search.

In MCTS we can classify positions into several types [11]: Those which are well suited to MCTS (good evaluation quality of the playouts), those which can only be solved by search (search-bound, the evaluation given by playouts is wrong but the search can resolve the problems in the position) and those which can only be solved by simulations (simulation-bound, the evaluation of the playouts is wrong and the search cannot resolve the problems, e.g., if there are several independent fights in Computer Go). The last type of position usually is difficult for MCTS because the playout policy is specified before the search and so knowledge of the policy determines if the position can be solved or not.

To reduce the impact of a fixed playout-policy, this paper describes a way to adapt the playout policy during a tree search by policy-gradient reinforcement learning [17]. This effectively improves the playout policy leading to better evaluations in the tree search. Positions which are simulation bound can be treated by adapting the policy so as to handle these positions correctly.

The contributions of our paper are as follows.

- We show a way to adapt playout policies inside MCTS by policy-gradient reinforcement learning.

© Springer International Publishing Switzerland 2015
A. Plaat et al. (Eds.): ACG 2015, LNCS 9525, pp. 1–11, 2015.
DOI: 10.1007/978-3-319-27992-3_1

- We show how to apply our approach to Computer Go.
- We conduct several experiments in Computer Go to evaluate the effect of our approach on the playing strength and the ability to solve simulation-bound positions.

The remainder of this paper is structured as follows.

In Sect. 2 we provide background information on policy-gradient reinforcement learning. In Sect. 3 we outline our approach to adapt playout polices in MCTS by policy-gradient reinforcement learning. In Sect. 4 we apply this algorithm to Computer Go and show the results of several experiments on playing strength and simulation-bound positions. In Sect. 5 we present related work. Finally, in Sect. 6 we draw our conclusion and point to future directions.

2   Background

In Sect. 3 we will use a policy gradient algorithm to optimize the strength of a policy inside MCTS. This section shortly derives the basics of the REINFORCE [18] algorithm. We use the *softmax policy* to generate playouts:

$$\pi_\theta(s, a) = \frac{e^{\phi(s,a)^T \theta}}{\sum_b e^{\phi(s,b)^T \theta}}. \tag{1}$$

The policy $\pi_\theta(s, a)$ specifies the probability of playing the move a in position s. It is calculated from the feature vector $\phi(s, a) \in \mathbb{R}^n$ and the vector $\theta \in \mathbb{R}^n$ of feature weights. For policy gradient algorithms we use the gradient of the log of the policy:

$$\nabla_\theta \log \pi_\theta(s, a) = \phi(s, a) - \sum_b \pi_\theta(s, b)\phi(s, b). \tag{2}$$

If a game is defined as a sequence of states and actions $g = (s_1, a_1, ..., s_n, a_n)$ and a result of the game $r(g)$ is 0 for a loss and 1 for a win, then the strength J of a policy π_θ is

$$J(\pi_\theta) = \sum_g p_{\pi_\theta}(g)r(g), \tag{3}$$

with the sum going over all possible games weighted by $p_{\pi_\theta}(g)$, the probability of their occurrence under policy π_θ. To optimize the strength of a policy the gradient of $J(\pi_\theta)$ can be calculated as follows.

$$\nabla_\theta J(\pi_\theta) = \sum_g \nabla_\theta p_{\pi_\theta}(g)r(g)$$

$$= \sum_g \nabla_\theta \Big(\prod_i \pi_\theta(s_i, a_i) \Big) r(g)$$

$$= \sum_g \Big(\prod_i \pi_\theta(s_i, a_i) \sum_j \frac{\nabla \pi_\theta(s_j, a_j)}{\pi_\theta(s_j, a_j)} \Big) r(g)$$

$$= \sum_g \left[p_{\pi_\theta}(g) r(g) \sum_j \frac{\nabla \pi_\theta(s_j, a_j)}{\pi_\theta(s_j, a_j)} \right]$$

$$= \sum_g \left[p_{\pi_\theta}(g) r(g) \sum_j \nabla \log \pi_\theta(s_j, a_j) \right]$$

$$= \mathbb{E} \left[r(g) \sum_j \nabla \log \pi_\theta(s_j, a_j) \right] \tag{4}$$

The last step is the expectation when games (the sequence of states/actions and the final result) are generated by the policy π_θ. We can approximate this gradient by sampling a finite number of games from the policy π_θ and then use this gradient to update the feature weights with stochastic gradient ascent. The resulting algorithm is called REINFORCE [18].

3   Adaptive Playouts

In MCTS we have a playout-policy $\pi(s, a)$ which specifies the probability distribution to be followed in the playouts. The policy is learned from expert games resulting in a fixed set of weights which is used during the tree search (offline learning). Here we assume that the playout policy is a softmax policy as defined in Eq. 1. To allow the playout policy to change during the tree search (online learning) we add additional features and weights to the policy:

$$\pi_\theta(s, a) = \frac{e^{\bar{\phi}(s,a)^T \bar{\theta} + \hat{\phi}(s,a)^T \hat{\theta}}}{\sum_b e^{\bar{\phi}(s,b)^T \bar{\theta} + \hat{\phi}(s,b)^T \hat{\theta}}}, \tag{5}$$

with $\bar{\phi}(s, a)$ and $\bar{\theta}$ the features and weights used during offline learning and $\hat{\phi}(s, a)$ and $\hat{\theta}$ for online learning. This difference between offline features which describe general knowledge and online features which can learn specific rules during a tree search is adapted from the Dyna-2 architecture which is used in reinforcement learning with value functions [16]. Offline feature weights are learned before the tree search and are kept fixed during the whole search. Online feature weights are used in addition to the offline feature weights and are set to zero before the search but can adapt during the search.

To leave the notation uncluttered, in the following we will only use the form of Eq. 1 instead of 5 whenever we talk about the policy in general and the distinction between online/offline features is clear from the context.

The overall change in MCTS is shown in Algorithm 1. During the search, for every playout a gradient has to be computed (line 11). After the playout has finished the policy parameters are changed with a stochastic gradient descent update (line 12) according to Eq. 8 which will be deduced in the following.

The policy-gradient update in line 12 performs the REINFORCE algorithm of the previous section. The overall aim of the policy-gradient algorithm is to optimize the strength of the policy. To avoid the policy to drift too much away from the offline policy we use a L2-regularization of the online-weights of the policy which keeps them close to zero. So, instead of maximizing the strength of the policy $J(\pi_\theta)$ we maximize:

Algorithm 1. MCTS with Adaptive Playouts

1: **function** MCTS(s, b)
2:     **if** $s \notin tree$ **then**
3:         expand(s)
4:     **end if**
5:     **if** totalCount(s) > 32 **then**
6:         $b \leftarrow$ winrate(s)
7:     **end if**
8:     $a \leftarrow$ select(s)
9:     $s' \leftarrow$ playMove(s, a)
10:     **if** expanded **then**
11:         $(r, g) \leftarrow$ playout(s')
12:         policyGradientUpdate(g, b)
13:     **else**
14:         $r \leftarrow$ MCTS(s', b)
15:     **end if**
16:     update(s, a, r)
17:     **return** r
18: **end function**

$$J(\pi_\theta) - \frac{\lambda}{2}\|\hat\theta\|^2. \tag{6}$$

Moreover, to stabilize the learning during the tree search we use a baseline b in the REINFORCE gradient [18]. This reduces the variance of the gradient but does not introduce bias:

$$(r - b)\sum_j \nabla \log \pi_\theta(s_j, a_j). \tag{7}$$

When traversing the tree down to a leaf in the selection part of MCTS, we set the baseline for the next playout to the winrate stored in the last tree node that occurred during this selection (line 6 in Algorithm 1). To ensure a stable estimate of the winrate only tree-nodes with at least 32 playouts are considered.

Summarizing, after a playout with a sequence of states and actions $(s_1, a_1, ..., s_n, a_n)$ in the MCTS with result r and the baseline b the policy is changed by a stochastic gradient-ascent step by

$$\hat\theta \leftarrow \hat\theta + \alpha \left[\left((r - b)\sum_j \nabla \log \pi_\theta(s_j, a_j) \right) - \lambda\hat\theta \right]. \tag{8}$$

In case of a two-player game we are adapting two policies (one for each player) simultaneously.

4 Experiments in Computer Go

In this section we outline how we applied adaptive playouts to 9×9 and 19×19 Computer Go. We first give an overview of the offline and online features used in the playouts. Then we conduct experiments on the playing strength of adaptive playouts and their impact on the speed of the playouts. Finally, we measured the effect of adaptive playouts on some difficult problems for MCTS programs in the so called "two-safe-groups" test set proposed in [11]. This test-set consists of simulation-bound positions, i.e., positions which are difficult for MCTS and can only be solved by a good playout policy in reasonable time.

4.1 Implementation

The core MCTS program used for the experiments makes use of state-of-the-art technologies like RAVE [8], progressive widening [12], progressive bias [7] and a large amount of knowledge (shape and common fate graph patterns [9]) in the tree search part. On the internet server KGS, where computer programs can play against humans, with the inclusion of adaptive playouts it has reached a rank of 3 dan under the name ABAKUS.

For the policy-gradient update of Eq. 8 we choose the L2-regularization parameter $\lambda = 0.001$ and learning-rate schedule of $\alpha_t = \frac{\alpha_0}{1+\alpha_0 \lambda t}$ as recommended in [5] with t the number of iterations already done in a Monte-Carlo Tree Search and $\alpha_0 = 0.01$. Adaptive weights are set to zero on each new game but are carried over to successive tree searches in the same game.

4.2 Features

Features of ABAKUS are similar to those mentioned in [10]. The playout policy $\pi(s, a)$ is based on 3×3 patterns around the move a and local features based on the last move played. The 3×3 patterns are extended with atari information (1 liberty or > 1 liberty) of the 4 direct neighbours and are position independent. All local features are related to the last move.

1. Contiguous to the last move.
2. Save new atari-string by capturing.
3. Save new atari-string by capturing but resulting in self-atari.
4. Save new atari-string by extending.
5. Save new atari-string by extending but resulting in self-atari.
6. Solve ko by capturing.
7. 2-point semeai: if the last move reduces a string to two liberties any move which kills a neighboring string with 2 liberties has this feature.
8. 3-4-5-point semeai heuristic: if the last move reduces a string to 3,4 or 5 liberties and this string cannot increase its liberties (by a move on its own liberties) then any move on a liberty of a neighboring string of the opponent which also cannot increase its liberties, has this feature.

9. Nakade capture: if the last move captured a group with a nakade shape the killing move has this feature.

These features represent the offline features $\bar{\phi}$ and their corresponding weights are learned before any tree search from expert games. To incorporate online features $\hat{\phi}$ into the playout policy the same features are used in a position-dependent way, i.e., for each intersection on the Go board we have features for every 3×3 pattern and every local feature. To decrease the number of features the 3×3 patterns are hashed to 32 different values. This does produce a large number of hashing conflicts but in practice this can be neglected as in a concrete position usually only a few patterns occur. The advantage is a large increase in performance as the weight vector is smaller. For the contiguous feature we also add the information of where the last move was.

4.3 Speed

We measured the speed of MCTS with and without adaptive playouts on a dual-socket Intel Xeon E5-2670 (total 16 cores), 2.6 GHz and 64 GByte main memory. On an empty board we can run 1337 playouts/s with static playouts and 1126 playouts/s with adaptive playouts per thread. Therefore, using adaptive playouts reduces the overall speed of the program to about 84 %.

In parallel MCTS we share the adaptive weights between all threads and synchronize writes with a lock. This has the advantage that all threads adapt the same weights and thus can learn faster. On the other hand this imposes a lock-penalty on the program when synchronizing. This penalty can be seen in Fig. 1 where we measure the number of playouts per second depending on the number of parallel threads. While the performance ratio of adaptive to static playouts is 84 % for a single thread it decreases to 68 % for 16 threads.

4.4 Playing Strength

To measure the improvement in playing strength by using adaptive playouts we played several tournaments by Abakus against the open source program Pachi [4] on the 19×19 board with 7.5 komi. Each tournament consisted of 1024 games

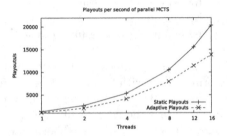

Fig. 1. Performance of parallel MCTS

Table 1. Playing strength of ABAKUS against PACHI, 1024 games played for each entry, 95 % confidence intervals

Thinking time (ABAKUS/PACHI)	Winrate static	ELO static	Winrate adaptive	ELO adaptive
8,000/32,000 playouts	48.4 % ± 3.1	−11	66.0 % ± 2.9	+115
16,000/32,000 playouts	63.4 % ± 3.0	+95	75.3 % ± 2.6	+193
32,000/32,000 playouts	72.9 % ± 2.7	+171	86.1 % ± 2.1	+317
One second both	62.0 % ± 3.0	+85	69.9 % ± 2.8	+146
Two seconds both	53.6 % ± 3.1	+25	68.9 % ± 2.8	+138
Four seconds both	48.7 % ± 3.1	−9	68.8 % ± 2.8	+137

with each program running on a 16 core Xeon E5-2670 with 2.6 GHz. Pondering was turned off for both programs.

The results are shown in Table 1. PACHI was set to a fixed number of 32,000 playouts/move while ABAKUS with static and adaptive playouts used 8,000, 16,000 and 32,000 playouts/move. The table shows a large improvement of adaptive playouts against static playouts of roughly 15 %. In this setting ABAKUS with 8,000 playouts/move is about as strong as PACHI with 32,000 playouts, with equal playouts/move ABAKUS beats PACHI in 86 % of the games.

As ABAKUS in general is slower than PACHI (mainly due to more knowledge in the tree and a different type of playout policy) and as shown in the previous section the adaptive playouts decrease the speed of playouts considerable, we also conducted experiments with equal time for both programs. In Table 1 you see the results for one, two and four seconds thinking time for both programs. In this setting PACHI runs about 27,000 playouts per second on an empty board, ABAKUS with static playouts 20,000 and with adaptive playouts 13,000. Despite the slowdown with adaptive playouts we see an increase in playing strength of about 130 ELO beating PACHI in 68 % of the games. Moreover, adaptive playouts show a much better scaling than static playouts. The adaptive playouts keep a winrate of about 68 % for all time settings. In contrast, the static playouts start with a winrate of 62 % with one second thinking time but decrease towards 48 % when the thinking time is increased to four seconds.

4.5 Two Safe Groups

The two-safe-groups test set was created in [11] to show the limits of MCTS. In all 15 positions white is winning with two groups on the board which are alive. The problem for MCTS is that both groups can easily die in the playouts leading to a wrong winrate estimation of the playouts (simulation-bound positions). As all positions are won by white the winrate at the root position of the MCTS should be close to zero. In the figures there is a target reference line of 30 % winrate (for black) and all test-cases should converge below this line. One example position of the testcases can be seen in Fig. 2.

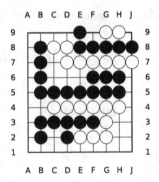

Fig. 2. Test case two

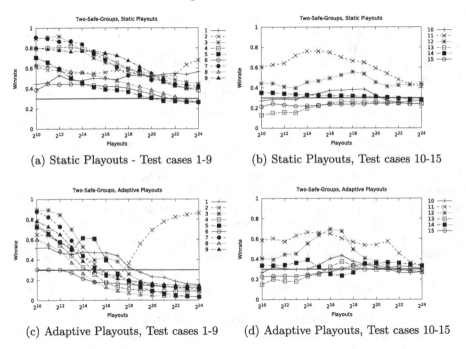

(a) Static Playouts - Test cases 1-9

(b) Static Playouts, Test cases 10-15

(c) Adaptive Playouts, Test cases 1-9

(d) Adaptive Playouts, Test cases 10-15

Fig. 3. Results of the two-safe-groups test cases

In Fig. 3(a) and (b) we see the results of MCTS with static playouts. If the evaluation at the beginning is wrong (high winrate for black) MCTS has difficulties to converge to the correct evaluation staying above the 30 % line. While in theory MCTS converges to the correct evaluation the test positions contain too deep fights to be resolved by plain MCTS in reasonable time.

Figure 3(c) shows a subset of all test cases with adaptive playouts where the playouts can fully adapt to the given positions. The positions are still evaluated wrong at the beginning but once the playouts have adapted to them MCTS behaves normal and slowly converges to the correct solution. Figure 3(d) shows the result for adaptive playouts and the test-cases where the playout policy could not adapt to the positions optimally. In all of them one of the groups cannot be resolved by the playouts (due to missing features). Still the program can find the correct solution but it takes a large amount of time for the winning rate to converge towards zero.

The strength of MCTS in combination with adaptive playouts can be seen in test case two (Fig. 2). While this position was created to be a win for white instead it is a win for black (black can attack the bottom white group by j4 as confirmed by Shih-Chieh Huang, the author of these test-cases; an example line is black j4-j3-j2-j5-j6 and black wins the resulting ko-fight). To play the correct moves a program first has to understand that both groups are alive by normal play which can be seen in Fig. 3(c). In test-case two the winrate drops until 2^{17} playouts. Up to there the program has adapted the playout policy to understand that both groups are alive. Only then the search can spot the correct solution by attacking the bottom white group with j4 and the winrate starts to increase to over 80 %.

5 Related Work

Much of our work is inspired by David Silver's work. In his RLGO [15] a value function is learned by temporal difference learning for every position in the search. This value function is then used by an ϵ-greedy policy. In this way the policy adapts to the positions encountered in the search. He later combined this TD-Search with the conventional tree-search algorithms alpha-beta and MCTS. He first executed a TD-Search to learn the value function and the used this knowledge in the tree-search algorithms. While this improved alpha-beta, in MCTS only knowledge learned from offline games improved the search. The knowledge gained through the TD-Search could not improve MCTS.

In our work we use the same difference of offline and online knowledge but instead of learning a value function we learn the playout policy directly with policy-gradient reinforcement learning. Moreover, we incorporate the learning directly into the MCTS (without a two-phase distinction) so that the policy adapts to the same positions we encounter in the search.

Hendrik Baier introduced adaptive playouts based on the Last-Good-Reply (LGR) policy [1]. Whenever a playout is won by, e.g., black, for every move from white the response of black is stored in a table. In later playouts black can then replay successful responses according to the table. In an extension called LGRF, Baier and Drake implemented that if a playout is lost all move-answers of the playout are removed from the table [2]. If there are no moves in the move-answer-table then a default policy is followed. The LGRF policy improved the Go program OREGO considerably but discussions on the computer go mailing list [3] showed that it had no positive effect on other programs like PACHI or OAKFOAM.

Evolutionary adaptation of the playout policy was used in [13] and later successfully applied to general video game playing [14]. They adapted the playout policy inside MCTS with the help of an evolutionary algorithm. The main difference to our work is the type of optimization algorithm to improve the playout policy. While in [13, 14] an evolutionary algorithm was used, in our approach we use policy-gradient reinforcement learning.

6   Conclusions and Future Work

This paper shows a way how to adapt playout policies in Monte-Carlo Tree Search. To achieve this we use policy-gradient reinforcement learning which directly optimizes the strength of the policy. We outlined how to adapt policies in general and applied it to the concrete case of Computer Go. The experiments indicate that the adaptation of policies results in a large increase of playing strength but also a decrease in speed. As a net result the program ABAKUS still achieved an increase of about 130 ELO in playing strength. Moreover, ABAKUS was able to solve several instances of the "two-safe-groups" test set which is known to be very difficult for MCTS programs.

Future work includes optimizing the performance of adaptive playouts. The experiments showed that parallel MCTS is harmed by a global lock on the adaptive weights. Removing this bottleneck will lead to a direct increase in playing strength and allows scaling to more than 16 threads.

Additionally, the playout policies need informative features to learn useful moves. The experiments showed that simple and fast features can adapt to quite complex cases. Nevertheless, it is an open question which features are best suited to improve adaptive playouts. In contrast to hand-made features, non-parametric features which are built during the learning process or even neural networks could be a promising direction.

Finally, there is a huge amount of choice in reinforcement-learning algorithms. In this paper we explored the use of a simple policy-gradient algorithm. More complex Actor-Critic algorithms could improve adaptive playouts further.

References

1. Baier, H.: Adaptive playout policies for Monte Carlo go. Master's thesis, Osnabrueck University, Germany (2010)
2. Baier, H., Drake, P.: The power of forgetting: improving the last-good-reply policy in Monte Carlo go. IEEE Trans. Comput. Intell. AI Games 2(4), 303–309 (2010)
3. Baudis, P.: Effect of LGRF on the playing strength agains gnugo. Website 15 June 2012. http://www.mail-archive.com/computer-go@dvandva.org/msg05060.html. Accessed 09 March 2015
4. Baudiš, P., Gailly, J.: PACHI: state of the art open source go program. In: van den Herik, H.J., Plaat, A. (eds.) ACG 2011. LNCS, vol. 7168, pp. 24–38. Springer, Heidelberg (2012)
5. Bottou, L.: Stochastic gradient tricks. In: Montavon, G., Orr, G.B., Müller, K.-R. (eds.) Neural Networks, Tricks of the Trade, Reloaded. Lecture Notes in Computer Science, vol. 7700, pp. 430–445. Springer, Heidelberg (2012)

6. Browne, C., Powley, E., Whitehouse, D., Lucas, S., Cowling, P., Rohlfshagen, P., Tavener, S., Perez, D., Samothrakis, S., Colton, S.: A survey of Monte Carlo tree search methods. IEEE Trans. Comput. Intell. AI Games **4**(1), 1–43 (2012)
7. Chaslot, G., Winands, M., Uiterwijk, J., van den Herik, H., Bouzy, B.: Progressive strategies for Monte-Carlo tree search. New Math. Nat. Comput. **4**(3), 343–357 (2008)
8. Gelly, S., Silver, D.: Combining online and offline knowledge in UCT. In: Proceedings of the 24th International Conference on Machine Learning, ICML 2007, pp. 273–280, New York (2007)
9. Graf, T., Platzner, M.: Common fate graph patterns in Monte Carlo tree search for computer go. In: 2014 IEEE Conference on Computational Intelligence and Games (CIG), pp. 1–8, August 2014
10. Huang, S.-C., Coulom, R., Lin, S.-S.: Monte-carlo simulation balancing in practice. In: van den Herik, H.J., Iida, H., Plaat, A. (eds.) CG 2010. LNCS, vol. 6515, pp. 81–92. Springer, Heidelberg (2011)
11. Huang, S.-C., Müller, M.: Investigating the limits of Monte-Carlo tree search methods in computer go. In: Herik, H.J., Iida, H., Plaat, A. (eds.) CG 2013. LNCS, vol. 8427, pp. 39–48. Springer, Heidelberg (2014)
12. Ikeda, K., Viennot, S.: Efficiency of static knowledge bias in Monte-Carlo tree search. In: Herik, H.J., Iida, H., Plaat, A. (eds.) CG 2013. LNCS, vol. 8427, pp. 26–38. Springer, Heidelberg (2014)
13. Lucas, S.M., Samothrakis, S., Pérez, D.: Fast evolutionary adaptationfor Monte Carlo tree search. In: Esparcia-Alcázar, A.I., Mora, A.M. (eds.) EvoApplications 2014. LNCS, vol. 8602, pp. 349–360. Springer, Heidelberg (2014)
14. Perez, D., Samothrakis, S., Lucas, S.: Knowledge-based fast evolutionary MCTS for general video game playing. In: 2014 IEEE Conference on Computational Intelligence and Games (CIG), pp. 1–8, August 2014
15. Silver, D.: Reinforcement learning and simulation-based search in computer go. Ph.D. thesis, University of Alberta (2009)
16. Silver, D., Sutton, R.S., Müller, M.: Sample-based learning and search with permanent and transient memories. In: Proceedings of the 25th International Conference on Machine Learning, ICML 2008, pp. 968–975 (2008)
17. Szepesvari, C.: Algorithms for Reinforcment Learning. Morgan and Claypool, USA (2010)
18. Williams, R.J.: Simple statistical gradient-following algorithms for connectionist reinforcement learning. Mach. Learn. **8**, 229–256 (1992)

Early Playout Termination in MCTS

Richard Lorentz[✉]

Department of Computer Science, California State University,
Northridge, CA 91330-8281, USA
lorentz@csun.edu

Abstract. Many researchers view mini-max and MCTS-based searches
as competing and incompatible approaches. For example, it is generally
agreed that chess and checkers require a mini-max approach while Go
and Havannah require MCTS. However, a hybrid technique is possible
that has features of both mini-max and MCTS. It works by stopping
the random MCTS playouts early and using an evaluation function to
determine the winner of the playout. We call this algorithm MCTS-EPT
(MCTS with early playout termination) and study it using MCTS-EPT
programs we have written for Amazons, Havannah, and Breakthrough.

1 Introduction

Monte-Carlo Tree Search (MCTS) differs from "classical" mini-max game-tree
search in two major ways. First, no evaluation function is needed in MCTS.
Instead, the random playouts in the MCTS act as a kind of sampling of the
possible outcomes from various board positions, which in turn can be used to
rate (evaluate) these different positions. Also, MCTS builds the search tree so
that more promising lines of play are more thoroughly explored in the tree than
less promising ones. As a result, we have learned that MCTS can drastically
outperform mini-max based search engines in games where evaluation functions
are difficult to obtain, and especially in games with large branching factors [1,2].

A hybrid approach to MCTS is possible, however. Instead of allowing the
random playout to run until the end of the game we can instead terminate the
playout early and then apply an evaluation function to the position to determine
which side is likely to win. We call this approach MCTS with early playout
termination (MCTS-EPT, or simply EPT). A number of successful programs
have been written using EPT. See, for example, [5–7,9].

We have written EPT programs that play the games of Amazons, Breakthrough, and Havannah and we will refer to them as AMABOT, BREAKBOT,
and HAVBOT. AMABOT was originally written using mini-max techniques and,
playing under the name INVADER, was one of the top Amazons programs at
the Computer Olympiads from 2001–2005 [11], but never finished in first place.
After converting from mini-max to EPT, AMABOT has won each of the last five
Computer Olympiads it has entered.

BREAKBOT is a more recent program. In contrast to AMABOT, it was originally written as an MCTS program and then was migrated over to the MCTS-
EPT approach. The pure MCTS version played a fairly average game, whereas

© Springer International Publishing Switzerland 2015
A. Plaat et al. (Eds.): ACG 2015, LNCS 9525, pp. 12–19, 2015.
DOI: 10.1007/978-3-319-27992-3_2

the EPT incarnation is very strong, being one of the top 3 players on the Little Golem game-playing Web site [10], where it plays under the name WANDERER.

HAVBOT was also originally pure MCTS [8] that has recently been converted to EPT. HAVBOT was also a moderately strong MCTS program, but is only slightly stronger using EPT. Like BREAKBOT, HAVBOT also plays under the name WANDERER. It has played in a number of Computer Olympiads and also plays on the Little Golem Web site.

Though creating an EPT program is straightforward, we will explain in detail (1) the requirements and difficulties of producing a strong EPT program from the perspective of our success with AMABOT and BREAKBOT and (2) our difficulties with HAVBOT.

2   History

We begin with a brief history of our research into MCTS-EPT. By 2007 AMABOT had performed well in a number of Computer Olympiads, but had never managed to win one. Johan de Koning's program, 8QP, was the five-time winner of the event and we could not seem to reach its level of play. Also in 2007 the MCTS revolution was in full swing, so we wondered what MCTS could offer us beyond what our mini-max program was providing. The mini-max program was using a sophisticated evaluation function so we had little hope that MCTS would be able to achieve the same level of play without using all the knowledge that was available to the evaluation function. Unknown to us at the time, Julien Kloetzer was doing the same research under the guidance of Hiroyuki Iida [5]. As it turns out we independently came to the same conclusion, namely, random playouts were insufficient. We needed to use the large amount of knowledge that was coded in the evaluation function. We also both discovered that the evaluation function can best be used as EPT rather than, say, to help guide the random playouts. In the case of AMABOT, we were ultimately able to achieve a win rate of 80 % using EPT over the mini-max based program. We then went on to win the next five Computer Olympiads using EPT. We believe Kloetzer's program had the potential for similar results, but he did not have the luxury of a pre-existing, strong evaluation function, leaving him at a disadvantage.

In 2009, motivated in part by the "Havannah Challenge" [12], a number of projects began to develop an Havannah playing program, including our HAVBOT project. Havannah seemed to be a perfect candidate for MCTS because of its high move branching factor, its very large state space, and the fact that a good evaluation function seems very hard to find. With but one exception, all known Havannah playing programs use MCTS. The one exception is a mini-max based program written by the talented game programmer Johan de Koning. The one time he entered it in the Computer Olympiad it lost every game it played against the other two programs, providing strong evidence that MCTS is the approach of choice.

However, progress in Havannah programming has not progressed as we might have hoped. Though the top programs do play at a reasonable level, about the level of somebody who has played the game for 6 months or a year, they still play with a very unnatural style, and often win their games by virtue of tactical shots missed by the human opponent. Our feeling is that Havannah programs cannot be expected to play at an elite level until they learn to play a more natural, human-like game. Towards this end, we have retooled HAVBOT to use EPT. Evidence is still inconclusive, and more details will be provided below, but we feel that its current style of play is more natural and has the potential to improve to noticeably higher levels of play. It currently beats the mini-max version of HAVBOT about 60 % of the time.

BREAKBOT, like HAVBOT, was written initially using MCTS but we fully expected to transition to EPT. As was the case with the MCTS version of AMABOT, without an evaluation function it's level of play languished in the low intermediate range. With the introduction of EPT it's level rose considerably and quickly, where after quite a bit of work, it is, at the time of this writing, the third highest rated player on Little Golem, and the second highest rated active player. The evidence that BREAKBOT with EPT outperforms MCTS is convincing. What is not quite so obvious is if it is better than mini-max based programs. The evidence we have to support this viewpoint is that there are two other programs playing on Little Golem, both of them are mini-max based, and BREAKBOT has won the majority of the encounters, though against the stronger of the two, LUFFYBOT, all of the games have been very close. We may conclude that EPT stands up well against mini-max and even though many of the games have been close, BREAKBOT ultimately outperforms the mini-max based ones.

3 Details

We now consider implementation details for MCTS-EPT. Our conclusions concerning these details are drawn from many years of experimenting (beginning in 2007) with the three different programs across two different playing situations (real time as played in the Computer Olympiads and very slow as played on the turn-based Web site Little Golem). As such some features seem to span most EPT situations while others apply to more specific settings.

3.1 Blending Mini-Max and EPT

It would seem natural that certain phases of a game would lend themselves to mini-max analysis while others to EPT. In fact, for many years AMABOT was written so that EPT was used throughout the majority of the game, and then switched over to mini-max near the end. Evidence seemed to indicate that the breadth-first nature of mini-max was superior near the end of the game because it would be less likely to miss a tactical shot that EPT (and MCTS in general) might miss because EPT had gotten stuck on a "good" line of play and did not have the time to find a better move. This, of course, is a general problem with MCTS, and can be fatal near the end game when a missed winning line or a failed proper defence can quickly and permanently turn a game around.

We now believe this is not true for two reasons. First, it is easy to incorporate solvers into MCTS, and therefore EPT, by propagating wins and losses up the MCTS tree in the usual and/or fashion. The advantage of being able to prove nodes outweighs anything lost by the tendency of MCTS to get stuck on a suboptimal line of play. Further, the solver can accelerate the exit from a bad line of play because winning and losing positions propagate immediately up the tree rather than requiring many simulations to reach the same conclusion.

Secondly, it is simply the case that the strengths of MCTS extend well to all aspects of the game. A good example is seen when dealing with defective territory in Amazons, a problem that turns up near the end of the game. This has always been a bit of a sticky issue with programs because the overhead necessary to deal with defects, typically done either by using patterns or other computationally expensive procedures in the evaluation function, does not seem to be worth the cost. In the case of EPT, however, defective territory is easily detected. In the presence of defective territory the MCTS tree accurately assesses the defect because the random playouts show that the territory cannot be properly filled. As a result, AMABOT has not used any mini-max and has been exclusively an EPT program for the last 5 years.

3.2 Progressive Widening

It is usually necessary to assist EPT by focusing on promising moves above and beyond what the MCTS rules suggest. In nodes with very few visits it can be difficult to distinguish among the many children. Two apparently different techniques have been developed that accomplish essentially the same thing. Progressive widening restricts access to nodes with low evaluation values and low visit counts and gradually phases them in as the parent node gets more visits [4]. Alternatively, node initialization (sometimes referred to as *priors* [3]) initializes the win and visit counts of nodes at the time of their creation with win values that reflect the strength of the node, again determined by the evaluation function.

In all three of our programs we have seen that it is necessary to use one of these techniques. In the case of AMABOT, progressive widening is used. In fact, since AMABOT possesses such a mature and accurate evaluation function and since Amazons allows so many legal moves, especially in the early parts of the game, we push progressive widening a bit further and do some forward pruning. Amazons positions can have more than 2000 legal moves. When building the EPT tree, we evaluate all possible children of a node and only put the top 750 in the tree and then from these we proceed with the usual progressive widening.

With HAVBOT and BREAKBOT we use the evaluation function to initialize win values in new nodes. Considerable tuning is necessary to find good initial values because, as is so common with MCTS related algorithms, we must find the proper balance so that the tree grows without inappropriate bias. In all three cases the winning advantage when using these techniques is significant, being over 75 %.

3.3   When to Terminate

Certainly a fundamental question is: when should the playout be terminated. The longer we delay the termination the more the behavior is like pure MCTS while sooner terminations put added emphasis on the evaluation function. We were surprised to find that in all three of our programs the optimal termination point was quite early and nearly at the same point in all three cases, namely, after around five moves. When the evaluation function is known to be quite reliable, as is the case with AMABOT, and to a lesser extent BREAKBOT, it is not too surprising that an earlier termination should be preferred since additional random playouts before evaluating will only dilute the effect of the evaluation. However, in the case of HAVBOT, where the evaluation is still very much a work in progress and can be quite undependable, the optimal termination point is still about the same and later termination points degrade its behavior at a rate quite similar to what is observed in AMABOT. In essence, it appears that even a weak evaluation function can compare favorably with a long random playout.

But what about termination points that are shorter than the optimal value? Since all three programs show similar results, let us focus on BREAKBOT. Though it stands to reason that shorter termination points might outperform longer ones when these termination points are reasonably large, it is not immediately obvious why the optimal value is not 1 or 0.

Consider Fig. 1 where we show the results of BREAKBOT playing as white against 4 other versions that were modified to have different termination points. Terminating after four random moves is optimal. Delaying the termination point beyond the optimal quickly degrades the performance and it is a bit surprising just how quickly it degrades.

Termination	Winning result
1	33%
4	43%
6	27%
12	10%

Fig. 1. Playout termination points in BREAKBOT.

But of particular interest is the first row that clearly shows that only 1 random move is not as good as 4. The values for 2 and 3 random moves degraded roughly uniformly. Why is it the case that a few random moves actually improve performance?

To help us understand this phenomenon we ran hundreds of games where at every position a move was generated by two versions of BREAKBOT, with termination points of 4 and 1. We found that on average the different versions disagreed on the best move about 12 times per game, where the average length of a game is 55 moves. It is important to point out, however, that a similar test performed on two copies of the same version of BREAKBOT (with termination point 4) still disagreed an average of 7 times per game, simply because of the random nature of EPT. In general, this suggests that about 5 times a game, or roughly 10 % of the time, the termination-1 version selects a move that the termination-4 version probably would not, and presumably more often than not this is a weaker move. Visual examination of these moves, however, generally does not reveal major blunders. Rather, when differences are detectable at all, they are small and subtle. Of course, five minor mistakes a game is certainly sufficient to cause the observed drop in winning percentage.

But it is difficult to provide a definitive explanation as to exactly what causes these roughly five aberrations per game. Why would fewer random moves in a playout hinder performance? Observational evidence suggests it boils down to a trade off between the advantages of a deep evaluation and disadvantages of losing information from the randomness of a playout. In general, an evaluation near the end of the game is more reliable than one earlier on but after too many random moves a position may lose the essence of the starting position. We search for a happy medium where a few random moves take us closer to the end of the game, without having the random moves degrade the information too much. For all three games we are studying this cutoff seems to be around 4 or 5.

Related to this, we should mention the concept of improving the random playouts. This, of course, is an important technique for an MCTS program and is certainly one of the major reasons MCTS programs are so successful. In the case of EPT it appears to be of little or no help. On the one hand it is not too surprising given that the random playouts are only 4 or 5 moves deep, but on the other hand given how important it is for MCTS, we thought we could get some improvement in our programs by improving the playouts. Despite considerable effort on all three programs, we have never been able to demonstrate any advantage by introducing smart playouts.

Finally, we point out that in a game like Amazons the evaluation function can vary wildly depending on whose move it is. This is sometimes referred to as the parity effect. The evaluation function tends to heavily favor the player to move. To help stabilize EPT we can modify the random playouts so that they always terminate with the same player to move. In the case of Amazons we terminate the playout after either 5 or 4 moves, accordingly. This produces some small advantage in the case of AMABOT, but in the cases of HAVABOT and BREAKBOT, where the evaluations do not display such a strong parity effect, adjusting the playouts this way does not seem to have any effect.

3.4   Miscellaneous

In this section we summarize a few other observations and techniques that we consider important.

It is generally the case that MCTS programs prefer to record wins and losses at the end of their playouts, rather than trying to somehow keep track of the margin of victory. We find the same is true with EPT. Rather than somehow use the value of the evaluation function, we have always obtained the best results by simply treating the evaluation as a boolean function, reporting either a win or a loss.

In Sect. 3.1 mention was made of the fact that EPT, as well as MCTS, can get stuck on a bad move simply because there is not enough time for the refutation of this weak move to achieve sufficient visits. Even though this seems like a problem mainly for real-time play, we find the problem carries over even for turn-based play where we sometimes allow as much as 15 min of thinking time. This problem can occur anywhere in the tree, but we have found that if we deal with it specifically at the root, we can get better results. What we do is we increase the exploitation constant only at the root so that exploration is encouraged, allowing more moves to be considered. Specifically, since all of our EPT programs are UCT based, we simply increase the UCT constant, typically by a factor of around 6. It does not make sense to uniformly change the UCT constant by this amount because the constant has already been optimized. But changing it only at the root has the very real effect of possibly allowing a move to be considered that might otherwise have been ignored. We have not been able to prove an advantage quantitatively, but we have seen quite a few cases of games on Little Golem where moves were found that were clearly better than those found without the adjustment while the reverse has yet to be observed. This technique, as well as the next, would probably apply to MCTS programs as well.

In the case of BREAKBOT we had to deal with the situation that early captures are almost always bad. We found no satisfactory way to deal with this in the evaluation because a capture by the first player usually requires a recapture by its opponent, so it balances out in the evaluation. Attempts to recognize the bad exchange after the fact in the evaluation had too many undesirable side effects. Our solution was to deal with this in the move selection process of the MCTS part of the search. Whenever a move was being selected for traversal or expansion in the MCTS tree, if it was a capture we hand tuned a penalty value for its winning percentage. This penalty is a function of (1) the stage of the game (early, middle, or late) and (2) the depth in the tree in which the move is being considered. This proved to be a successful way to deal with a problem that we were unable to deal with in the evaluation.

4   Conclusions

We have had considerable success with MCTS-EPT in games with a variety of features. Amazons is a game with a fairly large branching factor but it does allow for very precise and sophisticated evaluation functions. Still, EPT AMABOT outperforms all mini-max based programs. Not only does AMABOT do well in real-time play but it has played a number of games against some of the strongest players on turn based Little Golem and has never lost.

Breakthrough has a smaller branching factor but evaluation functions tend to be rather primitive. Not many programs exist that play Breakthrough, but of the two we are aware of (both play on the Little Golem site), we know that both are mini-max based, and both have losing records to BREAKBOT.

Havannah is a game, like Go, that has no strong mini-max based programs and not until the MCTS revolution did any reasonable programs exist. The three strongest Havannah playing programs all play on Little Golem and, though maybe slightly weaker than the other two, the MCTS version of HAVBOT plays a very similar game to the other two. Even though the evaluation function for HAVBOT is still quite primitive the program is making some promising looking moves and is outperforming its MCTS counterpart. As the evaluation continues to improve we feel there is great potential for this program.

As a side note, AMABOT and BREAKBOT are now so strong that progress comes very slowly. When deciding if a modification is an improvement sometimes simply running test games is not sufficient. If the tests are inconclusive, we must be ready to allow human intervention. How do the moves look to us? Do they seem to improve on the older version moves? How often do the moves look worse? We must be willing to make decisions based on answers to these kinds of questions, especially in the setting of turn based play, where results come at an agonizingly slow pace.

References

1. Browne, C., Powley, D., Whitehouse, D., Lucas, S., Cowling, P., Rohlfshagen, P., Tavener, S., Perez, D., Samothrakis, S., Colton, C.: A survey of monte carlo tree search methods. IEEE Trans. Comput. Intell. AI Games **4**(1), 1–49 (2012)
2. Coulom, R.: Efficient selectivity and backup operators in monte-carlo tree search. In: 5th International Conference on Computers and Games, CG 2006, Turin, Italy, pp. 72–84 (2006)
3. Gelly, S., Silver, D.: Combining online and offline knowledge in UCT. In: Ghahramani, Z. (ed.) Proceedings of the 24th International Conference on Machine Learning (ICML 2007), pp. 273–280. ACM, New York (2007)
4. Chaslot, G.M.J.-B., Winands, M.H.M., van den Herik, H.J., Uiterwijk, J.W.H.M., Bouzy, B.: Progressive strategies for monte-carlo tree search. New Math. Nat. Comput. **4**(3), 343–357 (2008)
5. Kloetzer, J., Iida, H., Bouzy, B.: The monte-carlo approach in amazons. In: Computer Games Workshop, Amsterdam, The Netherlands, pp. 113–124 (2007)
6. Lorentz, R.J.: Amazons discover monte-carlo. In: van den Herik, H.J., Xu, X., Ma, Z., Winands, M.H.M. (eds.) CG 2008. LNCS, vol. 5131, pp. 13–24. Springer, Heidelberg (2008)
7. Lorentz, R., Horey, T.: Programming breakthrough. In: van den Herik, H.J., Iida, H., Plaat, A. (eds.) CG 2013. LNCS, vol. 8427, pp. 49–59. Springer, Heidelberg (2013)
8. Lorentz, R.: Experiments with monte-carlo tree search in the game of havannah. ICGA J. **34**(3), 140–150 (2011)
9. Winands, M.H.M., Björnsson, Y.: Evaluation function based monte-carlo LOA. In: van den Herik, H.J., Spronck, P. (eds.) ACG 2009. LNCS, vol. 6048, pp. 33–44. Springer, Heidelberg (2010)
10. http://www.littlegolem.net/jsp/index.jsp
11. http://www.grappa.univ-lille3.fr/icga/program.php?id=249
12. Havannah#The Havannah Challenge. https://chessprogramming.wikispaces.com/

Playout Policy Adaptation for Games

Tristan Cazenave[✉]

LAMSADE, Université Paris-Dauphine, Paris, France
cazenave@lamsade.dauphine.fr

Abstract. Monte-Carlo Tree Search (MCTS) is the state of the art algorithm for General Game Playing (GGP). We propose to learn a playout policy online so as to improve MCTS for GGP. We test the resulting algorithm named Playout Policy Adaptation (PPA) on ATARIGO, BREAKTHROUGH, MISERE BREAKTHROUGH, DOMINEERING, MISERE DOMINEE-RING, GO, KNIGHTTHROUGH, MISERE KNIGHTTHROUGH, NOGO and MISERE NOGO. For most of these games, PPA is better than UCT with a uniform random playout policy, with the notable exceptions of Go and Nogo.

1 Introduction

Monte-Carlo Tree Search (MCTS) has been successfully applied to many games and problems [2]. The most popular MCTS algorithm is Upper Confidence bounds for Trees (UCT) [17]. MCTS is particularly successful in the game of Go [7]. It is also the state of the art in HEX [15] and General Game Playing (GGP) [10,20]. GGP can be traced back to the seminal work of Jacques Pitrat [21]. Since 2005 an annual GGP competition is organized by Stanford at AAAI [14]. Since 2007 all the winners of the competition use MCTS.

Offline learning of playout policies has given good results in Go [8,16] and HEX [15], learning fixed pattern weights so as to bias the playouts.

The RAVE algorithm [13] performs online learning of moves values in order to bias the choice of moves in the UCT tree. RAVE has been very successful in Go and HEX. A development of RAVE is to use the RAVE values to choose moves in the playouts using Pool RAVE [23]. Pool RAVE improves slightly on random playouts in HAVANNAH and reaches 62.7 % against random playouts in Go.

Move-Average Sampling Technique (MAST) is a technique used in the GGP program CADIA PLAYER so as to bias the playouts with statistics on moves [10,11]. It consists of choosing a move in the playout proportionally to the exponential of its mean. MAST keeps the average result of each action over all simulations. Moves found to be good on average, independent of a game state, will get higher values. In the playout step, the action selections are biased towards selecting such moves. This is done using the Gibbs (or Boltzmann) distribution. Playout Policy Adaptation (PPA) also uses Gibbs sampling. However, the evaluation of an action for PPA is not its mean over all simulations such as in MAST. Instead the value of an action is learned comparing it to the other available actions for the states where it has been played.

© Springer International Publishing Switzerland 2015
A. Plaat et al. (Eds.): ACG 2015, LNCS 9525, pp. 20–28, 2015.
DOI: 10.1007/978-3-319-27992-3_3

Later improvements of CADIA PLAYER are N-Grams and the last good reply policy [27]. They have been applied to GGP so as to improve playouts by learning move sequences. A recent development in GGP is to have multiple playout strategies and to choose the one which is the most adapted to the problem at hand [26].

A related domain is the learning of playout policies in single-player problems. Nested Monte-Carlo Search (NMCS) [3] is an algorithm that works well for puzzles. It biases its playouts using lower level playouts. At level zero NMCS adopts a uniform random playout policy. Online learning of playout strategies combined with NMCS has given good results on optimization problems [22]. Online learning of a playout policy in the context of nested searches has been further developed for puzzles and optimization with Nested Rollout Policy Adaptation (NRPA) [24]. NRPA has found new world records in Morpion Solitaire and crosswords puzzles. The principle is to adapt the playout policy so as to learn the best sequence of moves found so far at each level. PPA is inspired by NRPA since it learns a playout policy in a related fashion and adopts a similar playout policy. However, PPA is different from NRPA in multiple ways. NRPA is not suited for two-player games since it memorizes the best playout and learns all the moves of the best playout. The best playout is ill-defined for two-player games since the result of a playout is either won or lost. Moreover a playout which is good for one player is bad for the other player so learning all the moves of a playout does not make much sense. To overcome these difficulties PPA does not memorize a best playout and does not use nested levels of search. Instead of learning the best playout it learns the moves of every playout but only for the winner of the playout.

NMCS has been previously successfully adapted to two-player games in a recent work [5]. PPA is a follow-up to this paper since it is the adaptation of NRPA to two-player games.

In the context of GGP we look for general enhancements of UCT that work for many games without game-specific tweakings. This is the case for PPA. In our experiments we use the exact same algorithm for all the games. It is usually more difficult to find a general enhancement than a game specific one. PPA is an online learning algorithm, it starts from scratch for every position and learns a position specific playout policy each time.

We now give the outline of the paper. The next section details the PPA algorithm and particularly the playout strategy and the adaptation of the policy. The third section gives experimental results for various games, various board sizes and various numbers of playouts. The last section concludes.

2   Online Policy Learning

PPA is UCT with an adaptive playout policy. It means that it develops a tree exactly as UCT does. The difference with UCT is that in the playouts PPA has a weight for each possible move and chooses randomly between possible moves proportionally to the exponential of the weight.

In the beginning PPA starts with a uniform playout policy. All the weights are set to zero. Then, after each playout, it adapts the policy of the winner of the playout. The principle is the same as the adaptation of NRPA except that it only adapts the policy of the winner of the playout with the moves of the winner.

The PPA-playout algorithm is given in Algorithm 1. It takes as parameters the board, the next player, and the playout policy. The playout policy is an array of real numbers that contains a number for each possible move. The only difference with a random playout is that it uses the policy to choose a move. Each move is associated to the exponential of its policy number and the move to play is chosen with a probability proportional to this value.

The PPA-adaptation algorithm is given in Algorithm 2. It is related to the adaptation algorithm of NRPA. The main difference is that it is adapted to games and only learns the moves of the winner of the playout. It does not use a best sequence to learn as in NRPA but learns a different playout every time. It takes as parameter the winner of the playout, the board as it was before the playout, the player to move on this board, the playout to learn and the current playout policy. It is parameterized by α which is the number to add to the weight of the move in the policy. The adapt algorithm plays the playout again and for each move of the winner it biases the policy towards playing this move. It increases the weight of the move and decreases the weight of the other possible moves on the current board.

The PPA algorithm is given in Algorithm 3. It starts with initializing the policy to a uniform policy containing only zeros for every move. Then it runs UCT for the given number of playouts. UCT uses the PPA-playout algorithm for its playouts. They are biased with the policy. The result of a call to the UCT function is one descent of the tree plus one PPA playout that gives the winner of this single playout. The playout and its winner are then used to adapt the policy using the PPA-adapt function. When all playouts have been played the PPA function returns the move that has the most playouts at the root as in usual UCT.

The UCT algorithm called by the PPA algorithm is given in Algorithm 4.

3   Experimental Results

We played PPA against UCT with random playouts. Both algorithms use the same number of playouts. The UCT constant is set to 0.4 for both algorithms as is usual in GGP. α is set to 1.0 for PPA. For each game we test two board sizes: 5×5 and 8×8, and two numbers of playouts: 1,000 and 10,000.

The games we have experimented with are:

- ATARIGO: the rules are the same as for GO except that the first player to capture a string has won. ATARIGO has been solved up to size 6×6 [1].
- BREAKTHROUGH: The game starts with two rows of pawns on each side of the board. Pawns can capture diagonally and go forward either vertically or diagonally. The first player to reach the opposite row has won. BREAKTHROUGH

Algorithm 1. The PPA-playout algorithm

---

playout (*board, player, policy*)
while true **do**
   if *board* is terminal **then**
      return winner (*board*)
   end if
   $z \leftarrow 0.0$
   for *m* in possible moves on *board* **do**
      $z \leftarrow z + \exp (policy [m])$
   end for
   choose a *move* for *player* with probability proportional to $\frac{exp(policy[move])}{z}$
   play (*board, move*)
   player $\leftarrow$ opponent (*player*)
end while

Algorithm 2. The PPA-adaptation algorithm

---

adapt (*winner, board, player, playout, policy*)
polp $\leftarrow$ *policy*
for *move* in *playout* **do**
   if *winner* = *player* **then**
      polp [*move*] $\leftarrow$ *polp* [*move*] $+ \alpha$
      $z \leftarrow 0.0$
      for *m* in possible moves on *board* **do**
         $z \leftarrow z + \exp (policy [\text{m}])$
      end for
      for *m* in possible moves on *board* **do**
         polp [*m*] $\leftarrow$ *polp* [*m*] - $\alpha * \frac{exp(policy[m])}{z}$
      end for
   end if
   play (*board, move*)
   player $\leftarrow$ opponent (*player*)
end for
policy $\leftarrow$ *polp*

Algorithm 3. The PPA algorithm

PPA (*board, player*)
for i in 0, maximum index of a move **do**
   policy[*i*] $\leftarrow$ 0.0
end for
for i in 0, number of playouts **do**
   b $\leftarrow$ *board*
   winner $\leftarrow$ UCT (*b, player, policy*)
   b1 $\leftarrow$ *board*
   adapt (*winner, b1, player, b.playout, policy*)
end for
return the move with the most playouts

Algorithm 4. The UCT algorithm

UCT (*board, player, policy*)
moves ← possible moves on *board*
if *board* is terminal **then**
    return winner (*board*)
end if
t ← entry of *board* in the transposition table
if *t* exists **then**
    bestValue ← −∞
    for *m* in *moves* **do**
        t ← *t.totalPlayouts*
        w ← *t.wins*[*m*]
        p ← *t.playouts*[*m*]
        value ← $\frac{w}{p} + c \times \sqrt{\frac{log(t)}{p}}$
        if *value* > *bestValue* **then**
            bestValue ← *value*
            bestMove ← *m*
        end if
    end for
    play (*board, bestMove*)
    player ← opponent (*player*)
    res ← UCT (*board, player, policy*)
    update *t* with *res*
else
    t ← new entry of *board* in the transposition table
    res ← playout (*board, player, policy*)
    update *t* with *res*
end if
return *res*

---

has been solved up to size 6 × 5 using Job Level Proof Number Search [25]. The best program for Breakthrough 8 × 8 uses MCTS combined with an evaluation function after a short playout [19].

- Misere Breakthrough: The rules are the same as for Breakthrough except that the first player to reach the opposite row has lost.
- Domineering: The game starts with an empty board. One player places dominoes vertically on the board and the other player places dominoes horizontally. The first player that cannot play has lost. Domineering was invented by Göran Andersson [12]. Jos Uiterwijk recently proposed a knowledge based method that can solve large rectangular boards without any search [28].
- Misere Domineering: The rules are the same as for Domineering except that the first player that cannot play has won.
- Go: The game starts with an empty grid. Players alternatively place black and white stones on the intersections. A completely surrounded string of stones is removed from the board. The score of a player at the end of a game with

Table 1. Win rate against UCT with the same number of playouts as PPA for various games of various board sizes using either 1,000 or 10,000 playouts per move.

	Size	Playouts	
		1,000	10,000
ATARIGO	5 × 5	81.2	90.6
ATARIGO	8 × 8	72.2	94.4
BREAKTHROUGH	5 × 5	60.0	56.2
BREAKTHROUGH	8 × 8	55.2	54.4
MISERE BREAKTHROUGH	5 × 5	95.0	99.6
MISERE BREAKTHROUGH	8 × 8	99.2	97.8
DOMINEERING	5 × 5	62.6	50.0
DOMINEERING	8 × 8	48.4	58.0
MISERE DOMINEERING	5 × 5	63.4	62.2
MISERE DOMINEERING	8 × 8	76.4	83.4
GO	5 × 5	21.2	23.6
GO	8 × 8	23.0	1.2
KNIGHTTHROUGH	5 × 5	42.4	30.2
KNIGHTTHROUGH	8 × 8	64.2	64.6
MISERE KNIGHTTHROUGH	5 × 5	95.8	99.8
MISERE KNIGHTTHROUGH	8 × 8	99.8	100.0
NOGO	5 × 5	61.8	71.0
NOGO	8 × 8	64.8	46.4
MISERE NOGO	5 × 5	66.4	67.8
MISERE NOGO	8 × 8	80.6	89.4

chinese rules is the number of her[1] stones on the board plus the number of her eyes. The player with the greatest score has won. We use a komi of 7.5 for white. GO was the first tremendously successful application of MCTS to games [7–9,18]. All the best current GO programs use MCTS.

- KNIGHTTHROUGH: The rules are similar to BREAKTHROUGH except that the pawns are replaced by knights that can only go forward.
- MISERE KNIGHTTHROUGH: The rules are the same as for KNIGHTTHROUGH except that the first player to reach the opposite row has lost.
- NOGO: The rules are the same as GO except that it is forbidden to capture and to commit suicide. The first player that cannot move has lost. There exist computer NOGO competitions and the best players use MCTS [4,6,9].
- MISERE NOGO: The rules are the same as for NOGO except that first player that cannot move has won.

---

[1] For brevity, we use 'he' and 'his', whenever 'he or she' and 'his or her' are meant.

We do not give results for single-player games since PPA is tailored to multi-player games. Also we do not compare with NMCS and NRPA since these algorithms are tailored to single-player games and perform poorly when applied directly to two-player games. We give results of 1,000 and 10,000 playouts per move.

Results are given in Table 1. Each result is the outcome of a 500 games match, 250 playing first and 250 playing second.

PPA has worse results than UCT in three games: Go, KNIGHTTHROUGH 5×5 and NOGO 8 × 8. For the other 17 games it improves over UCT. It is particularly good at misere games, a possible explanation is that it learns to avoid losing moves in the playouts and that it may be important for misere games that are waiting games.

We observe that PPA scales well in ATARIGO, MISERE BREAKTHROUGH, MISERE DOMINEERING, KNIGHTTHROUGH, MISERE KNIGHTTHROUGH and MISERE NOGO: it is equally good or even better when the size of the board or the number of playouts is increased. On the contrary it does not scale for Go and NOGO.

A possible explanation of the bad behaviour in Go could be that moves in Go can be either good or bad depending on the context and that learning an overall evaluation of a move can be misleading.

In the context of GGP, the time used by GGP programs is dominated by the generation of the possible moves and by the calculation of the next state. So biasing the playout policy is relatively unexpensive compared to the time used for the interpretation of the rules of the game.

4   Conclusion

In the context of GGP we presented PPA, an algorithm that learns a playout policy online. It was tested on ten different games for increasing board sizes and increasing numbers of playouts. On many games it scales well with board size and number of playouts and it is better than UCT for 33 out of the 40 experiments we performed. It is particularly good at misere games, scoring as high as 100 % against UCT at MISERE KNIGHTTHROUGH 8 × 8 with 10,000 playouts.

PPA is tightly connected to the NRPA algorithm for single-player games. The main differences with NRPA are that it does not use nested levels nor a best sequence to learn. Instead it learns the moves of each playout for the winner of the playout.

Future work include combining PPA with the numerous enhancements of UCT. Some of them may be redundant but others will probably be cumulative. For example combining PPA with RAVE could yield substantial benefits in some games.

A second line of research is understanding why PPA is good at many games and bad at other games such as Go. It would be interesting being able to tell the features of a game that make PPA useful.

References

1. Boissac, F., Cazenave, T.: De nouvelles heuristiques de recherche appliquées à la résolution d'Atarigo. In: Intelligence artificielle et jeux, pp. 127–141. Hermes Science (2006)
2. Browne, C., Powley, E., Whitehouse, D., Lucas, S., Cowling, P., Rohlfshagen, P., Tavener, S., Perez, D., Samothrakis, S., Colton, S.: A survey of Monte Carlo tree search methods. IEEE Trans. Comput. Intell. AI Games 4(1), 1–43 (2012)
3. Cazenave, T.: Nested Monte-Carlo search. In: Boutilier, C. (ed.) IJCAI, pp. 456–461 (2009)
4. Cazenave, T.: Sequential halving applied to trees. IEEE Trans. Comput. Intell. AI Games 7(1), 102–105 (2015)
5. Cazenave, T., Saffidine, A., Schofield, M., Thielscher, M.: Discounting and pruning for nested playouts in general game playing. GIGA at IJCAI (2015)
6. Chou, C.-W., Teytaud, O., Yen, S.-J.: Revisiting Monte-Carlo tree search on a normal form game: NoGo. In: Di Chio, C., et al. (eds.) EvoApplications 2011, Part I. LNCS, vol. 6624, pp. 73–82. Springer, Heidelberg (2011)
7. Coulom, R.: Efficient selectivity and backup operators in Monte-Carlo tree search. In: van den Herik, H.J., Ciancarini, P., Donkers, H.H.L.M.J. (eds.) CG 2006. LNCS, vol. 4630, pp. 72–83. Springer, Heidelberg (2007)
8. Coulom, R.: Computing elo ratings of move patterns in the game of go. ICGA J. 30(4), 198–208 (2007)
9. Enzenberger, M., Muller, M., Arneson, B., Segal, R.: Fuego - an open-source framework for board games and go engine based on Monte Carlo tree search. IEEE Trans. Comput. Intell. AI Games 2(4), 259–270 (2010)
10. Finnsson, H., Björnsson, Y.: Simulation-based approach to general game playing. In: AAAI, pp. 259–264 (2008)
11. Finnsson, H., Björnsson, Y.: Learning simulation control in general game-playing agents. In: AAAI (2010)
12. Gardner, M.: Mathematical games. Sci. Am. 230, 106–108 (1974)
13. Gelly, S., Silver, D.: Monte-Carlo tree search and rapid action value estimation in computer go. Artif. Intell. 175(11), 1856–1875 (2011)
14. Genesereth, M.R., Love, N., Pell, B.: General game playing: overview of the AAAI competition. AI Mag. 26(2), 62–72 (2005)
15. Huang, S., Arneson, B., Hayward, R.B., Müller, M., Pawlewicz, J.: Mohex 2.0: a pattern-based MCTS hex player. In: Computers and Games - 8th International Conference, CG 2013, Yokohama, Japan, 13–15 August 2013, Revised Selected Papers, pp. 60–71 (2013)
16. Huang, S.-C., Coulom, R., Lin, S.-S.: Monte-Carlo simulation balancing in practice. In: van den Herik, H.J., Iida, H., Plaat, A. (eds.) CG 2010. LNCS, vol. 6515, pp. 81–92. Springer, Heidelberg (2011)
17. Kocsis, L., Szepesvári, C.: Bandit based Monte-Carlo planning. In: Fürnkranz, J., Scheffer, T., Spiliopoulou, M. (eds.) ECML 2006. LNCS (LNAI), vol. 4212, pp. 282–293. Springer, Heidelberg (2006)
18. Lee, C., Wang, M., Chaslot, G., Hoock, J., Rimmel, A., Teytaud, O., Tsai, S., Hsu, S., Hong, T.: The computational intelligence of MoGo revealed in taiwan's computer go tournaments. IEEE Trans. Comput. Intell. AI Games 1(1), 73–89 (2009)
19. Lorentz, R., Horey, T.: Programming breakthrough. In: van den Herik, H.J., Iida, H., Plaat, A. (eds.) CG 2013. LNCS, vol. 8427, pp. 49–59. Springer, Heidelberg (2014)

20. Méhat, J., Cazenave, T.: A parallel general game player. KI **25**(1), 43–47 (2011)
21. Pitrat, J.: Realization of a general game-playing program. IFIP Congr. **2**, 1570–1574 (1968)
22. Rimmel, A., Teytaud, F., Cazenave, T.: Optimization of the nested Monte-Carlo algorithm on the traveling salesman problem with time windows. In: Di Chio, C., et al. (eds.) EvoApplications 2011, Part II. LNCS, vol. 6625, pp. 501–510. Springer, Heidelberg (2011)
23. Rimmel, A., Teytaud, F., Teytaud, O.: Biasing Monte-Carlo simulations through RAVE values. In: van den Herik, H.J., Iida, H., Plaat, A. (eds.) CG 2010. LNCS, vol. 6515, pp. 59–68. Springer, Heidelberg (2011)
24. Rosin, C.D.: Nested rollout policy adaptation for Monte Carlo tree search. In: IJCAI, pp. 649–654 (2011)
25. Saffidine, A., Jouandeau, N., Cazenave, T.: Solving BREAKTHROUGH with race patterns and job-level proof number search. In: van den Herik, H.J., Plaat, A. (eds.) ACG 2011. LNCS, vol. 7168, pp. 196–207. Springer, Heidelberg (2012)
26. Swiechowski, M., Mandziuk, J.: Self-adaptation of playing strategies in general game playing. IEEE Trans. Comput. Intell. AI Games **6**(4), 367–381 (2014)
27. Tak, M.J.W., Winands, M.H.M., Björnsson, Y.: N-grams and the last-good-reply policy applied in general game playing. IEEE Trans. Comput. Intell. AI Games **4**(2), 73–83 (2012)
28. Uiterwijk, J.W.H.M.: Perfectly solving domineering boards. In: Cazenave, T., Winands, M.H.M., Lida, H. (eds.) CGW 2013. Communications in Computer and Information Science, vol. 408, pp. 97–121. Springer, Switzerland (2013)

Strength Improvement and Analysis for an MCTS-Based Chinese Dark Chess Program

Chu-Hsuan Hsueh[1], I-Chen Wu[1(✉)], Wen-Jie Tseng[1], Shi-Jim Yen[2], and Jr-Chang Chen[3]

[1] Department of Computer Science, National Chiao Tung University, Hsinchu, Taiwan
{hsuehch,icwu,wenjie}@aigames.nctu.edu.tw
[2] Department of Computer Science and Information Engineering, National Dong Hwa University, Hualien, Taiwan
sjyen@mail.ndhu.edu.tw
[3] Department of Applied Mathematics, Chung Yuan Christian University, Taoyuan, Taiwan
jcchen@cycu.edu.tw

Abstract. Monte-Carlo tree search (MCTS) has been successfully applied to Chinese dark chess (CDC). In this paper, we study how to improve and analyze the playing strength of an MCTS-based CDC program, named DARKKNIGHT, which won the CDC tournament in the 17th Computer Olympiad. We incorporate the three recent techniques, early playout terminations, implicit minimax backups, and quality-based rewards, into the program. For early playout terminations, playouts end when reaching states with likely outcomes. Implicit minimax backups use heuristic evaluations to help guide selections of MCTS. Quality-based rewards adjust rewards based on online collected information. Our experiments showed that the win rates against the original DARKKNIGHT were 60.75 %, 70.90 % and 59.00 %, respectively for incorporating the three techniques. By incorporating all together, we obtained a win rate of 76.70 %.

1 Introduction

Chinese dark chess (CDC), widely played in Chinese community, is a two-player game and also a partially observable (PO) game with symmetric hidden information. The set of pieces in CDC is the same as those in Chinese chess; however, the pieces are faced down initially so both players do not know what the pieces are, until they are flipped.

In [8], the state space complexity of the game was estimated to be 10^{37} between those of Draughts and chess, while the game-tree complexity was estimated to be 10^{135} between those of chess and Chinese chess without considering chance nodes. In [33], Yen *et al.* estimated the game-tree complexity with chance nodes to be 10^{207} between those of Chinese chess and Shogi.

In the past, many CDC game-playing programs were developed. In the early stage, most CDC programs [8, 27] were developed using *alpha-beta search*. Recently, Yen *et al.* [33] incorporated *Monte-Carlo tree search (MCTS)* into a CDC program, named DIABLO, which won the following four tournaments [20, 26, 32, 35]. Ones of the authors of this paper also implemented an MCTS-based CDC program, named DARKKNIGHT, which won two CDC tournaments [28, 34], including that in the 17th Computer Olympiad.

© Springer International Publishing Switzerland 2015
A. Plaat et al. (Eds.): ACG 2015, LNCS 9525, pp. 29–40, 2015.
DOI: 10.1007/978-3-319-27992-3_4

This paper incorporates three recent techniques into DARKKNIGHT and analyzes how well these techniques improve the game-playing strength. These techniques are as follows.

1. *Early playout terminations* [2, 12, 19, 21–23, 30, 31]: Terminate a playout much earlier when it is very likely to win, lose, or draw.
2. *Implicit minimax backups* [19]: Guide MCTS selections by using heuristic evaluations of tree nodes together with the original simulated win rates.
3. *Quality-based rewards* [24]: Use simulation length and terminal state quality to adjust the rewards returning from simulations.

Our experiments showed that implicit minimax backups improved the most, which reached a win rate of 70.90 % against the original DARKKNIGHT, serving as the baseline program. The improvements reached a win rate of 60.75 % for early playout terminations, and 57.50 % and 59.00 % respectively when using simulation length and terminal state quality for quality-based rewards. By incorporating all together, we successfully improved DARKKNIGHT with a win rate of 76.70 %.

This paper is organized as follows. Section 2 reviews the game CDC and the previous work for CDC game-playing programs, and briefly introduces the MCTS algorithm. Section 3 presents the above three techniques, including the reviews of these techniques and the incorporations into DARKKNIGHT. Section 4 shows the experimental results. Finally, Sect. 5 makes concluding remarks.

2 Background

2.1 CDC

CDC [8, 33] is a two-player zero-sum non-deterministic game played on a 4×8 square board as illustrated in Fig. 1(a) and (b). The two players, *Red* and *Black*, respectively own identical sets of sixteen pieces with different colors. The piece types are shown in Fig. 1(c). Each piece has two faces, one showing the piece type and the other showing the piece cover which is identical for all pieces. When the piece type faces up, the type is revealed and known by both players. When it faces down, it is unrevealed and unknown.

Two kinds of actions are allowed in CDC: *flipping* and *moving*. Each flipping action flips a piece, namely making the piece type revealed. Each moving action is to move a revealed piece by the player owning the piece. All pieces can be moved to empty neighboring squares (with one square up, down, left or right). Pieces except cannons can be moved to neighboring squares with capturing the opponent's pieces that have equal or lower ranks shown in Fig. 1(c) with the following exceptions: Pawns can capture the opponent's king, but not the other way around. Cannons have a different capturing rule called *one-step-jump* in which any opponent's pieces can be captured as described in more details in [8, 33].

Initially, all 32 pieces are unrevealed and placed randomly, as in Fig. 1(a). So, the probability distributions of unrevealed pieces are all equal. The first player flips one of the 32 pieces and then owns the set of pieces of the revealed color, while the second

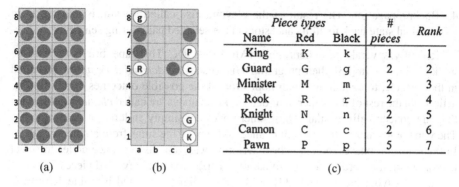

Fig. 1. (a) Initial board, (b) a board example and (c) piece information

player owns the other set. A player wins by capturing all of the opponent's pieces or making the opponent have no legal moves. The game also ends with a draw when both players play without any capturing and flipping within 40 plies, or when the same position appears three times.

2.2 Previous Work for CDC Game-Playing Programs

In the early stage, most CDC programs [8, 27] were developed using alpha-beta search. Chen *et al.* published a paper [8] about alpha-beta search in their CDC program, FLIPPER. Some more research work for CDC includes opening books [7], endgame databases [9, 10, 25], solving smaller CDC [6] and game variations [15].

Recently, MCTS has been incorporated into CDC game-playing programs [16, 17, 33]. MCTS [5] is a best-first search algorithm on top of a search tree, named *UCT* [18], using Monte-Carlo simulations as state evaluations. It has been successfully applied to several games, such as Go [11, 13, 14], General Game Playing (GGP) [3], Backgammon [29] and Phantom-Go [4]. There are four phases in MCTS:

1. Selection. A path is traversed from the root to one of the leaves following a selection policy. A popular selection policy is based on an *upper confidence bounds* function [1]:

$$\text{argmax}_i \left\{ Q_i + C\sqrt{\frac{\ln N}{n_i}} \right\} \tag{1}$$

where Q_i is an estimator for the value of node i, n_i the visit count of node i, N the visit count of the parent of node i and C a constant representing the weight of exploration. Commonly, Q_i is a win rate, w_i/n_i, or a mean value for rewards, where w_i is the win count of node i.

2. Expansion. One or more children are expanded from the selected leaf.
3. Playout. Following a playout policy, the game is played from the selected leaf to some terminal state.

4. Backpropagation. The result of the playout, also called reward, is updated from terminal state back to all its ancestors in the selected path during selection.

Recently, several researchers applied MCTS to CDC. This paper briefly reviews the work in [33, 34]. In [33], their program DIABLO used *non-deterministic nodes*, so-called in their article, to contain all *inner nodes* for all the possible outcomes of the flipping actions. In the rest of this paper, non-deterministic nodes are called *chance nodes*, since they are conceptually similar. Thus, inner nodes are simply children of chance nodes. The win counts and visit counts of chance nodes were the sums from all their children. In the selection phase, roulette wheel selection was applied to chance nodes to select one of children according to the probability distribution of unrevealed pieces.

In the playout phase, they [33] used piece weights, as listed in Table 1, in their program DIABLO when capturing pieces. The program DARKKNIGHT [34] was developed independently, but it used almost the same weights as in [33], except for reducing them by a factor of 100. The program DARKKNIGHT serves as the baseline program in this paper.

Table 1. The piece weights

Piece type	K/k	G/g	M/m	R/r	N/n	C/c	P/p
Weights in [33]	5500	5000	2500	1000	800	3000	800
Weights in this paper	55	50	25	10	8	30	8

3 Incorporated Techniques

Below we discuss the three incorporated techniques: early playout terminations in Subsect. (3.1), implicit minimax backups in Subsect. (3.2), and quality-based rewards in Subsect. (3.3).

3.1 Early Playout Terminations

For CDC, it is important to terminate playouts earlier for two reasons, speedup and accuracy of simulated results. The first reason is obvious. The second is that it is more accurate to terminate playouts at the time when they are very likely to win, lose, or draw. For example, it is normally a draw in a situation, Gmm, where Red has only one guard and Black has two ministers. In the case of keeping playing in the playout, Black may lose accidentally, while it is supposed to be a draw since G cannot capture either m. Thus, an early termination is actually more accurate than a continuous playout. Another example is KGmm. No matter where the pieces are placed and whether the pieces are revealed or not, Red always wins the game, since both K and G can capture m, but not *vice versa*. However, it may result in draws in playouts since Red may not be able to go to capture black pieces tactically during playouts.

In the rest of this subsection, we first review the previous work about early playout terminations and then describes our work to incorporate it into DARKKNIGHT.

Previous Work. In [12], a method was proposed to collect information online, and terminate playouts when the game obviously favors some player from the collected information. They used it to save the time for more simulations for GGP.

Some other researchers were to return results after making a fixed number of moves in playouts. In [21–23], for Amazons, Breakthrough and Havannah, they used evaluation functions to determine the winners who the evaluated scores favor. In [2], they did a shallow minimax search with fixed depths and then returned the evaluated scores directly. Their method was tested for Othello, Breakthrough and Catch the Lion.

In [30, 31], their methods for Lines of Action (LOA) were to check the scores by evaluation function every three moves, and then return the results in the following cases. If the scores exceeded some threshold, the results were wins. On the contrary, if the scores were below some threshold, the results were losses. Checking evaluation functions every three moves was for the sake of performance.

Our Work. Playouts are terminated earlier when detecting a *likely outcome* which is *win*, *loss*, or *draw*, from Red's perspective. The detection rules are based on a material combination [9, 25], namely a set of remaining pieces on the board in CDC. For CDC, the total number of legal material combinations is 8,503,055 ($= (2^1 3^5 6^1)^2 - 1$). In our work, we analyzed all the combinations according to some heuristic rules. From the analysis, 750,174 are set to win, 750,174 loss, 108,136 draw, and the rest are unknown. For example, the likely outcomes for KGmm, KGk and KGggm are set to win, and those for Gmm, CCccp and KPggm are draw. Note that we call it a *likely* outcome since the outcome may not be always true in a few extreme cases as illustrated by the following example. For KGggm, Black can try to exchange one g with G, though it is hard to do so. If Black successfully exchanges g with G without losing m, then the game becomes a draw.

In playouts, if the likely outcome for a material combination is one of win, loss and draw, the playouts end immediately and return the outcomes. The overhead for the detection is little, since we can simply lookup a table to check this. So, the detection is done for every move in playouts.

3.2 Implicit Minimax Backups

This subsection reviews the previous work for implicit minimax backups and then describes our work to incorporate it into the baseline program.

Previous Work. In [19], the researchers proposed implicit minimax backups which incorporated *heuristic evaluations* into MCTS. The heuristic evaluations were used together with the estimator Q_i to guide the selections of MCTS. Namely, Q_i in Formula (1) was replaced by the following:

$$(1 - \alpha)Q_i + \alpha m_i \tag{2}$$

where m_i is the minimax score of node i by heuristic evaluation and α is the weight of the minimax score.

In each leaf of UCT, a heuristic evaluation function was applied to compute the score, scaled to [−1, 1] through a sigmoid function. Then, the scaled score was also backed up as classical minimax search. For consistency, Q_i was ranged in [−1, 1], too.

They improved the playing strength of the following three games, Kalah, Breakthrough and LOA, and obtained win rates of around 60 % to 70 % in different settings of the self-play games. Their work also showed that even simple evaluation functions can lead to better performance. However, they found that this technique did not work on every game. They ran experiments on Chinese checkers and the card game Heart but obtained no significant improvement.

Our Work. For CDC, we use as the heuristic evaluation function the weighted material sum, namely the difference of the total piece weights between the player and the opponent. The piece weights are listed in Table 1. The heuristic evaluations are then scaled to [−1, 1] by a sigmoid function.

One issue to discuss is the heuristic evaluations for chance nodes, which were not mentioned in [19]. An intuitive way is to use the probability distribution for unrevealed pieces, when calculating the expected values for chance nodes. For example, suppose to have four unrevealed pieces, two Ps, one k and one m (as in Fig. 1(b)). The probability for P is 1/2, while those for the other two are 1/4. However, a problem occurs when unrevealed pieces are not in the UCT yet. For example, if k has not been flipped yet, the corresponding heuristic evaluation is unknown, making it hard to get the heuristic evaluations for chance nodes.

In order to solve this problem, we use the ratios of visit counts as the probabilities for all children of chance nodes. For the above example, assume to flip P 3 times and m once, but none for k. Then, we use 3/4 for P and 1/4 for m.

3.3 Quality-Based Rewards

This subsection reviews the previous work for quality-based rewards and then describes our work to incorporate it into the baseline program.

Previous Work. In [24], the researchers proposed new measurements to adjust the rewards obtained from simulations. Simulation length and terminal state quality were used as *quality assessments* of simulations to adjust the rewards from wins and losses.

In [24], simulation length, a domain independent quality assessment, was defined as the length of simulated game from the root to the terminal state in the simulation. Intuitively, in an advantageous position, the shorter the simulation length is, the better.

This technique maintained online sample mean and sample standard deviation of simulation length, denoted by μ_L^p and σ_L^p, for player p when p wins. With these statistical values, the reward R_L for simulation length was calculated as follows:

$$R_L = R + sign(R) \times a \times b(\lambda_L) \tag{3}$$

where R is the reward (e.g., 1 for a win and −1 for a loss), a the influence for this quality, $b(\lambda)$ a sigmoid function as shown in Formula (4) which scales the values to [−1, 1], λ_L a normalized value in Formula (5),

$$b(\lambda) = -1 + \frac{2}{1 + e^{-k\lambda}} \tag{4}$$

$$\lambda_L = \frac{\mu_L^p - 1}{\sigma_L^p} \tag{5}$$

k a constant to be tuned and l is the length of that simulation.

For terminal state quality, a function describing quality of terminal states was needed. For example, for Breakthrough, the piece difference between the winning and losing players was used to measure the terminal state quality, which was scaled to [0, 1]. Let μ_T^p and σ_T^p respectively denote sample mean and sample standard deviation of terminal state quality for player p. Similarly, the rewards R_T for terminal state quality was calculated in a way similar to Formula (3) as follows.

$$R_T = R + sign(R) \times a \times b(\lambda_T) \tag{6}$$

where λ_T is a normalized value in Formula (7),

$$\lambda_T = \frac{t - \mu_T^p}{\sigma_T^p} \tag{7}$$

and t is the terminal state quality. In Formula (3) and (6), a is a constant or is calculated according to the data accumulated online. Their experimental results showed not much difference.

Our Work. We also incorporate into DARKKNIGHT the above two quality assessments, simulation length and terminal state quality, to adjust the rewards from wins and losses. For draws, the simulations are not sampled. The two quality assessments are measured when simulations end. Without early playout terminations, MCTS simulations end when one player wins. With early playout terminations, MCTS simulations also end when one obtains a likely outcome with win or loss.

At terminal states, simulation length is simply the same as the one described in [24], and terminal state quality is obtained in the following way. First, simply count the remaining pieces of the winner. Then, incorporate domain knowledge like piece weights. Namely, we use the following formula:

$$\sum_{i \in S} (1 + c_w \times \omega_i) = |S| + c_w \sum_{i \in S} \omega_i \tag{8}$$

where S is the set of remaining pieces of the winner, $|S|$ the size of S, c_w a coefficient and ω_i the weight of piece i as in Table 1. The larger the coefficient c_w is, the higher the influence of the piece weights. All values of terminal state quality are scaled to [−1, 1] according to a sigmoid function.

4   Experiments

In our experiments, each modified version played 1,000 games against the baseline, the original DARKKNIGHT. Among the 1,000 games, one played 500 games as the first and the other 500 as the second. For each game, one scored 1 point for a win, 0 for a loss and 0.5 for a draw. For a given version, the average score of 1,000 games was the win rate against the baseline.

Initially, we performed a strength test for the baseline with different numbers of simulations per move against the one with 10,000, as shown in Fig. 2. In the rest of experiments, unless explicitly specified, we chose the one with 30,000, which was reasonable in the sense of both strength and computation time. Namely, it had a win rate of 75.75 %, while it took about 3.2 min to run a game with one thread on machines equipped with Intel(R) Xeon(R) CPU E31225, 3.10 GHz.

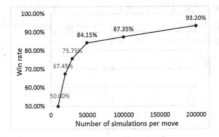

Fig. 2. Baseline with different numbers of simulations per move against that with 10,000

4.1 Experiments for Incorporating Individual Techniques

In this subsection, we incorporate techniques mentioned in Sect. 3 individually into the baseline to see how much they can improve.

Early Playout Terminations (EPT). As mentioned in Subsect. 3.1, we used material combinations to detect whether a playout reaches a terminal state earlier with a likely outcome, one of win, loss or draw. The experimental result showed a significant improvement with a win rate of 60.75 % against the baseline program. The result indicated that the accuracy of the simulated results was indeed increased with the help of the likely outcomes returned from EPT. In addition, the program with EPT also ran faster, at 32,000 simulations per second, than the one without EPT, at 27,000.

Implicit Minimax Backups (IMB). As mentioned in Subsect. 3.2, we used heuristic evaluations to help guide the selections of MCTS more accurately. In our experiments, we tested different weights of minimax score, α, and different numbers of simulations per move. The experimental results are shown in Fig. 3 which includes three lines, respectively, for 10,000, 30,000 and 100,000 simulations per move. For fairness, the corresponding baselines also ran the same numbers of simulations.

Figure 3 shows that the win rates are the highest when $\alpha = 0.3$ for the three lines, and 78.45 % for the one with 10,000 simulations per move, 70.90 % for 30,000 and 64.60 % for 100,000. The figure shows that IMB did significantly improve the playing strength. On the one hand, in the case that α was too high, the win rates went down for the following reason. The heuristic evaluations weighted too much higher than online

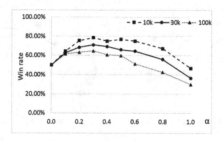

Fig. 3. Win rates for IMB with different α and different numbers of simulations per move

estimation which is usually more accurate than heuristic evaluations for a sufficiently large number of simulations. On the other hand, in the case that α was too low, the win rates also went down for the following reason. Less heuristic information was used to help guide the selections of MCTS accurately, since short-term tactical information provided by the minimax scores was missing, as explained in [19].

Figure 3 also has the following implication. For a higher number of simulations per move, the improvement was relatively smaller. This hints that IMB has less improvement for a sufficiently large number of simulations per move. The reason is: the help of minimax scores decreases, since simulated results become more accurate with more simulations.

In [19], they mentioned that IMB had no significant improvement for Heart, also a PO game. Interestingly, our results show that IMB did significantly improve the playing strength for CDC.

Quality-Based Rewards. As mentioned in Subsect. 3.3, we used simulation length (SL) and terminal state quality (TSQ) to adjust the rewards from simulations to favor those with shorter length and higher terminal state quality.

For SL, we tested two parameters, the influence, a, and the sigmoid function constant, k, and obtained the results as shown in Fig. 4(a). The highest win rate was 57.50 % when $a = 0.25$ and $k = 7$.

For TSQ, we needed one more parameter, c_w, a coefficient of piece weights for measuring terminal state quality. In our experiments, we obtained the highest win rate of 59.00 %, when $a = 0.3$, $k = 11$ and $c_w = 0.02$. By fixing $a = 0.3$ and $k = 11$, we obtained the highest at $c_w = 0.02$, slightly better than the one at $c_w = 0.01$, as in Fig. 4(b). Similarly, by fixing a and c_w, we obtained the highest at $k = 11$ as in Fig. 4(c). By fixing k and c_w, we obtained the highest win rate at $a = 0.3$ as shown in Fig. 4(d).

4.2 Combinations of Techniques

In this subsection, we further improved the playing strength by combining the above techniques, each of which used the best settings from the experimental results in the previous subsection. Table 2 first lists the best results for incorporating each individual technique, EPT, IMB, SL and TSQ (from the previous subsection). Then, we combined

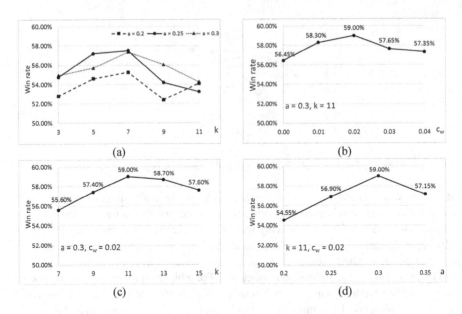

Fig. 4. The win rates for (a) SL, and for TSQ with different (b) c_w, (c) k and (d) a

both EPT and IMB first, denoted by EPT + IMB, since these two techniques improved the most when used alone. The result showed a win rate of 74.75 % for EPT + IMB.

Furthermore, we incorporated SL, TSQ, and SL + TSQ respectively into the above EPT + IMB. Their win rates are shown in Table 2. By incorporating all techniques together, the win rate reached up to 76.70 %.

Table 2. The best win rates of EPT, IMB, SL, TSQ and their combinations

	EPT	IMB	SL	TSQ	EPT + IMB	EPT + IMB + SL	EPT + IMB + TSQ	EPT + IMB + SL + TSQ
Win rate	60.75 %	70.90 %	57.50 %	59.00 %	74.75 %	76.00 %	75.95 %	76.70 %

5   Conclusion

To our best knowledge, this paper is the first attempt to incorporate the three techniques, early playout terminations (EPT), implicit minimax backups (IMB) and quality-based rewards (SL and TSQ) together and obtained significant improvement. We demonstrate this through an MCTS-based CDC game-playing program, DARKKNIGHT. Our experiments showed that all of these techniques did significantly improve the playing strength with win rates of 60.75 %, 70.90 %, 57.50 % and 59.00 % against the original DARKKNIGHT when incorporating EPT, IMB, SL and TSQ, respectively. By incorporating all together, the win rate reached up to 76.70 %. The results demonstrated the effectiveness of the above techniques for an MCTS-based CDC program. Besides, the enhanced DARKKNIGHT, with more fine tunings, won the CDC tournament in the 18th Computer Olympiad.

Acknowledgements. The authors would like to thank the Ministry of Science and Technology of the Republic of China (Taiwan) for financial support of this research under contract numbers MOST 102-2221-E-009-069-MY2, 102-2221-E-009-080-MY2, 104-2221-E-009-127-MY2, and 104-2221-E-009-074-MY2.

References

1. Auer, P., Cesa-Bianchi, N., Fischer, P.: Finite-time analysis of the multiarmed bandit problem. Mach. Learn. **47**(2–3), 235–256 (2002)
2. Baier, H., Winands, M.H.: Monte-Carlo tree search and minimax hybrids with heuristic evaluation functions. In: Cazenave, T., Winands, M.H., Björnsson, Y. (eds.) CGW 2014. CCIS, vol. 504, pp. 45–63. Springer, Heidelberg (2014)
3. Björnsson, Y., Finnsson, H.: CadiaPlayer: a simulation-based general game player. IEEE Trans. Comput. Intell. AI Games **1**(1), 4–15 (2009)
4. Borsboom, J., Saito, J.-T., Chaslot, G., Uiterwijk, J.: A comparison of Monte-Carlo methods for phantom go. In: Proceedings of BeNeLux Conference on Artificial Intelligence, Utrecht, The Netherlands, pp. 57–64 (2007)
5. Browne, C.B., Powley, E., Whitehouse, D., Lucas, S.M., Cowling, P.I., Rohlfshagen, P., Tavener, S., Perez, D., Samothrakis, S., Colton, S.: A survey of Monte Carlo tree search methods. IEEE Trans. Comput. Intell. AI Games **4**(1), 1–43 (2012)
6. Chang, H.-J., Hsu, T.-S.: A quantitative study of 2×4 Chinese dark chess. In: van den Herik, H., Iida, H., Plaat, A. (eds.) CG 2013. LNCS, vol. 8427, pp. 151–162. Springer, Heidelberg (2014)
7. Chen, B.-N., Hsu, T.-S.: Automatic generation of opening books for dark chess. In: van den Herik, H., Iida, H., Plaat, A. (eds.) CG 2013. LNCS, vol. 8427, pp. 221–232. Springer, Heidelberg (2014)
8. Chen, B.-N., Shen, B.-J., Hsu, T.-S.: Chinese dark chess. ICGA J. **33**(2), 93–106 (2010)
9. Chen, J.-C., Lin, T.-Y., Chen, B.-N., Hsu, T.-S.: Equivalence classes in chinese dark chess endgames. IEEE Trans. Comput. Intell. AI Games **7**(2), 109–122 (2015)
10. Chen, J.-C., Lin, T.-Y., Hsu, S.-C., Hsu, T.-S.: Design and implementation of computer Chinese dark chess endgame database. In: Proceeding of TCGA Workshop 2012, pp. 5–9, Hualien, Taiwan (2012) (in Chinese)
11. Enzenberger, M., Müller, M., Arneson, B., Segal, R.: Fuego: an open-source framework for board games and go engine based on Monte Carlo tree search. IEEE Trans. Comput. Intell. AI Games **2**(4), 259–270 (2010)
12. Finnsson, H.: Generalized Monte-Carlo tree search extensions for general game playing. In: The Twenty-Sixth AAAI Conference on Artificial Intelligence, pp. 1550–1556, Toronto, Canada (2012)
13. Gelly, S., Silver, D.: Monte-Carlo tree search and rapid action value estimation in computer go. Artif. Intell. **175**(11), 1856–1875 (2011)
14. Gelly, S., Wang, Y., Munos, R., Teytaud, O.: Modification of UCT with patterns in Monte-Carlo go. Technical report, HAL - CCSd - CNRS, France (2006)
15. Jouandeau, N.: Varying complexity in CHINESE DARK CHESS stochastic game. In: Proceeding of TCGA Workshop 2014, pp. 86, Taipei, Taiwan (2014)

16. Jouandeau, N., Cazenave, T.: Monte-Carlo tree reductions for stochastic games. In: Cheng, S.-M., Day, M.-Y. (eds.) TAAI 2014. LNCS, vol. 8916, pp. 228–238. Springer, Heidelberg (2014)

17. Jouandeau, N., Cazenave, T.: Small and large MCTS playouts applied to Chinese dark chess stochastic game. In: Cazenave, T., Winands, M.H., Björnsson, Y. (eds.) CGW 2014. CCIS, vol. 504, pp. 78–89. Springer, Heidelberg (2014)

18. Kocsis, L., Szepesvári, C.: Bandit based Monte-Carlo planning. In: Fürnkranz, J., Scheffer, T., Spiliopoulou, M. (eds.) ECML 2006. LNCS (LNAI), vol. 4212, pp. 282–293. Springer, Heidelberg (2006)

19. Lanctot, M., Winands, M.H.M., Pepels, T., Sturtevant, N.R.: Monte Carlo tree search with heuristic evaluations using implicit minimax backups. In: 2014 IEEE Conference on Computational Intelligence and Games, CIG 2014, pp. 1–8 (2014)

20. Lin, Y.-S., Wu, I.-C., Yen, S.-J.: TAAI 2011 computer-game tournaments. ICGA J. **34**(4), 248–250 (2011)

21. Lorentz, R.J.: Amazons discover Monte-Carlo. In: van den Herik, H., Xu, X., Ma, Z., Winands, M.H. (eds.) CG 2008. LNCS, vol. 5131, pp. 13–24. Springer, Heidelberg (2008)

22. Lorentz, R.: Early playout termination in MCTS. In: The 14th Conference on Advances in Computer Games (ACG2015), Leiden, The Netherlands (2015)

23. Lorentz, R., Horey, T.: Programming breakthrough. In: van den Herik, H., Iida, H., Plaat, A. (eds.) CG 2013. LNCS, vol. 8427, pp. 49–59. Springer, Heidelberg (2014)

24. Pepels, T., Tak, M.J., Lanctot, M., Winands, M.H.M.: Quality-based rewards for Monte-Carlo tree search simulations. In: 21st European Conference on Artificial Intelligence, Prague, Czech Republic (2014)

25. Saffidine, A., Jouandeau, N., Buron, C., Cazenave, T.: Material symmetry to partition endgame tables. In: van den Herik, H., Iida, H., Plaat, A. (eds.) CG 2013. LNCS, vol. 8427, pp. 187–198. Springer, Heidelberg (2014)

26. Su, T.-C., Yen, S.-J., Chen, J.-C., Wu, I.-C.: TAAI 2012 computer game tournaments. ICGA J. **37**(1), 33–35 (2014)

27. Theory of computer games, a course in National Taiwan University taught by Tsu, T.-S. http://www.iis.sinica.edu.tw/~tshsu/tcg/index.html

28. Tseng, W.-J., Chen, J.-C., Chen, L.-P., Yen, S.-J., Wu, I.-C.: TCGA 2013 computer game tournament report. ICGA J. **36**(3), 166–168 (2013)

29. Van Lishout, F., Chaslot, G., Uiterwijk, J.W.: Monte-Carlo tree search in Backgammon. In: Computer Games Workshop, pp. 175–184, Amsterdam, The Netherlands (2007)

30. Winands, M.H.M., Björnsson, Y., Saito, J.-T.: Monte Carlo tree search in lines of action. IEEE Trans. Comput. Intell. AI Games **2**(4), 239–250 (2010)

31. Winands, M.H., Björnsson, Y., Saito, J.-T.: Monte-Carlo tree search solver. In: van den Herik, H., Xu, X., Ma, Z., Winands, M.H. (eds.) CG 2008. LNCS, vol. 5131, pp. 25–36. Springer, Heidelberg (2008)

32. Yang, J.-K., Su, T.-C., Wu, I.-C.: TCGA 2012 computer game tournament report. ICGA J. **35**(3), 178–180 (2012)

33. Yen, S.-J., Chou, C.-W., Chen, J.-C., Wu, I.-C., Kao, K.-Y.: Design and implementation of Chinese dark chess programs. IEEE Trans. Comput. Intell. AI Games **7**(1), 66–74 (2015)

34. Yen, S.-J., Chen, J.-C., Chen, B.-N., Tseng, W.-J.: DarkKnight wins Chinese dark chess tournament. ICGA J. **36**(3), 175–176 (2013)

35. Yen, S.-J., Su, T.-C., Wu, I.-C.: The TCGA 2011 computer-games tournament. ICGA J. **34**(2), 108–110 (2011)

LinUCB Applied to Monte-Carlo Tree Search

Yusaku Mandai[✉] and Tomoyuki Kaneko

The Graduate School of Arts and Sciences, The University of Tokyo, Tokyo, Japan
mandai@graco.c.u-tokyo.ac.jp

Abstract. UCT is a de facto standard method for Monte-Carlo tree search (MCTS) algorithms, which have been applied to various domains and have achieved remarkable success. This study proposes a family of LinUCT algorithms that incorporate LinUCB into MCTS algorithms. LinUCB is a recently developed method that generalizes past episodes by ridge regression with feature vectors and rewards. LinUCB outperforms UCB1 in contextual multi-armed bandit problems. We introduce a straightforward application of LinUCB, LinUCT$_{PLAIN}$ by substituting UCB1 with LinUCB in UCT. We show that it does not work well owing to the minimax structure of game trees. To better handle such tree structures, we present LinUCT$_{RAVE}$ and LinUCT$_{FP}$ by further incorporating two existing techniques, rapid action value estimation (RAVE) and feature propagation, which recursively propagates the feature vector of a node to that of its parent. Experiments were conducted with a synthetic model, which is an extension of the standard incremental random tree model in which each node has a feature vector that represents the characteristics of the corresponding position. The experimental results indicate that LinUCT$_{RAVE}$, LinUCT$_{FP}$, and their combination LinUCT$_{RAVE-FP}$ outperform UCT, especially when the branching factor is relatively large.

1 Introduction

UCT [13] is a de facto standard algorithm of Monte-Carlo tree search (MCTS) [3]. UCT has achieved remarkable successes in various domains including the game of Go [8].

For game-tree search, UCT constructs and evaluates a game tree through a random sampling sequence. At each time step, a *playout* involving Monte-Carlo simulation is performed to improve the empirical estimation of the winning ratio at a leaf node and that of the ancestors of the leaf. For each playout, a leaf in the game tree is selected in a best-first manner by descending the most promising move with respect to the upper confidence bound of the winning ratio, UCB1 [1]. After it reaches the leaf, the score of the playout is determined by the terminal position, which is reached by alternatively playing random moves. Therefore, UCT works effectively without heuristic functions or domain knowledge. The fact is a remarkable advantage over traditional game-tree search methods based on alpha beta search [12], because such methods require adequate evaluation functions to estimate a winning probability of a position. Generally, the construction of such evaluation functions requires tremendous effort [11].

© Springer International Publishing Switzerland 2015
A. Plaat et al. (Eds.): ACG 2015, LNCS 9525, pp. 41–52, 2015.
DOI: 10.1007/978-3-319-27992-3_5

Although UCT does not explicitly require heuristics, many studies have incorporated domain-dependent knowledge into UCT to improve the convergence speed or playing strength, such as progressive widening [6], prior knowledge [10], and PUCB [16]. Such approaches utilize a set of features, i.e., a *feature vector*, which is observable in each state or in each move.

This study proposes a family of LinUCT as new MCTS algorithms. LinUCT is based on LinUCB [5,15], which has been studied in contextual multi-armed bandit problem [4]. While UCB1 only considers past rewards for each arm, LinUCB generalizes past episodes by ridge regression with feature vectors and rewards to predict the rewards of a future state. Thus, LinUCB is an alternative means to incorporate domain knowledge in an online manner. We first introduce LinUCT$_{PLAIN}$ by substituting UCB1 with LinUCB in UCT. However, this straightforward application of LinUCB is not promising, because it does not consider information in the subtree expanded under a node. To overcome this problem, we present LinUCT$_{RAVE}$ and LinUCT$_{FP}$, by incorporating two existing techniques, rapid action value estimation (RAVE) [9] and feature propagation that propagates feature vectors in a subtree to its root. We conducted experiments with a synthetic model that is a variant of incremental random trees that have served as good test sets for search algorithms [7,13,14,18,20]. We extend the trees such that each node has a feature vector while preserving the main property of the incremental random trees. The experiments demonstrate that LinUCT$_{RAVE}$, LinUCT$_{FP}$, and their combination LinUCT$_{RAVE-FP}$ outperform UCT, especially when the branching factor is relatively large.

2   MCTS and Algorithms in Multi-armed Bandit Problems

In the game of Go [2], minimax tree search does not work effectively because of the difficulties in constructing heuristic evaluation functions. After the emergence of a new paradigm, Monte-Carlo tree search (MCTS), the playing strength of computer players has been improved significantly in Go [8]. MCTS relies on a number of random simulations according to the following steps [3].

1. Selection: Starting at the root node, the algorithm descends the tree to a leaf. At each internal node, an algorithm for multi-armed bandit problems is employed to determine and select the move with the highest value with respect to a given criterion. Here, we consider UCB1, UCB1$_{RAVE}$, and LinUCB, as the selection criteria.
2. Expansion: The children of the leaf are expanded if appropriate, and one child is selected randomly.
3. Simulation: The rest of the game is played randomly. Typically, the game consists of completely random moves; however, some studies have suggested that a well-designed probability of moves yields improved performance [6,17].
4. Back propagation: The result (i.e., win/loss) of the simulation is back-propagated to the nodes on the path the algorithm descended in Step 1.

These steps are repeated until the computational budget (e.g., time) is exhausted. Then, the best move, typically the most visited child of the root, is identified.

2.1 UCT

UCT (UCB applied to trees) is one of the most successful MCTS variants. In the selection step at time t, UCT computes the UCB1 value [1] for move a of node s as follows:

$$\text{UCB1}(s, a) = \overline{X}_a + \sqrt{\frac{2\ln N_s}{N_a}} \tag{1}$$

where $\overline{X}_a$ is the empirical mean of the rewards of move a, and N_i is the number of visits to the node or move i. Note that a position after a move is defined without ambiguity in deterministic games. Thus, we use move a or the position after move a interchangeably for simplicity. The best move converges to the same move identified by a minimax algorithm under a certain assumption [13].

2.2 RAVE

RAVE is a remarkable enhancement to MCTS, particularly effective in Go [10]. It is a generalization of the All Moves As First (AMAF) heuristic, where AMAF treats all moves in a simulation as if they were selected as the first move. When RAVE is incorporated, the following interpolation UCB1$_{\text{RAVE}}$ is used as the selection criterion:

$$\text{UCB1}_{\text{RAVE}}(s, a) = \beta(s)\,\text{RAVE}(s, a) + (1 - \beta(s))\,\text{UCB1}(s, a) \tag{2}$$

where $\text{RAVE}(s, a)$ has a similar form as Eq. (1). Note that $\overline{X}_a$ and N_a are counted according to AMAF in Go. Thus, the RAVE value may become rapidly a good estimate of the reward by incorporating various simulation results that are not contained in UCB1 for $\overline{X}_a$. The interpolation weight $\beta(s)$ is $\sqrt{\frac{k}{3N_s + k}}$. Therefore, the value of UCB1$_{\text{RAVE}}$ converges to that of UCB1 for large N_s, while RAVE covers the unreliable prediction of UCB1 when N_s is small. Parameter k controls the number of episodes when both terms are equal [9].

2.3 LinUCB

LinUCB is an algorithm for the contextual bandit problems [5,15]. Here, a feature vector is observable for each arm, and it is assumed that the expected reward of arm a is defined by the inner product between the feature vector $\mathbf{x}_a$ and an unknown coefficient vector $\boldsymbol{\theta}_a^*$; $\mathbb{E}[r_a|\mathbf{x}_a] = \mathbf{x}_a^\top \boldsymbol{\theta}_a^*$, where r_a is the reward of arm a. The LinUCB algorithm employs ridge regression to estimate $\boldsymbol{\theta}_a^*$ using the trials performed so far. The criterion in arm selection in LinUCB is expressed as follows:

$$\text{LinUCB}(a) = \mathbf{x}_a^\top \hat{\boldsymbol{\theta}}_a + \alpha\sqrt{\mathbf{x}_a^\top \mathbf{A}_a^{-1}\mathbf{x}_a}, \tag{3}$$

Algorithm 1. LinUCB

Inputs: $\alpha \in \mathbb{R}_+$
for $t = 1, 2, 3, \ldots$ **do**
    for all $a \in \mathcal{A}_t$ **do**   ▷ $\mathcal{A}_t$ is a set of available arms at t
        if a is new **then**
            $\mathbf{A}_a \leftarrow \mathbf{I}_{d \times d}$ ▷ d dimensional identity matrix
            $\mathbf{b}_a \leftarrow \mathbf{0}_{d \times 1}$ ▷ d dimensional zero vector
        $\hat{\boldsymbol{\theta}}_a \leftarrow \mathbf{A}_a^{-1} \mathbf{b}_a$
        $p_a \leftarrow \mathbf{x}_a^\top \hat{\boldsymbol{\theta}}_a + \alpha \sqrt{\mathbf{x}_a^\top \mathbf{A}_a^{-1} \mathbf{x}_a}$
    $a_t \leftarrow \arg \max_{a \in \mathcal{A}_t} p_a$ with ties broken arbitrarily
    Observe a real-valued payoff r_t
    $\mathbf{A}_{a_t} \leftarrow \mathbf{A}_{a_t} + \mathbf{x}_{a_t} \mathbf{x}_{a_t}^\top$
    $\mathbf{b}_{a_t} \leftarrow \mathbf{b}_{a_t} + r_t \mathbf{x}_{a_t}$

---

where $\hat{\boldsymbol{\theta}}_a$ is the current estimate of $\boldsymbol{\theta}_a^*$, and $\mathbf{A}_a^{-1}$ is the inverse of the variance-covariance matrix for the regression on arm a. Constant $\alpha > 0$ controls the exploration-exploitation balance. With a probability of at least $1 - \delta$, the difference between the current estimate $\mathbf{x}_a^\top \hat{\boldsymbol{\theta}}_a$ and the expected reward $\mathbb{E}[r_a | \mathbf{x}_a]$ is bounded [15, 19] as follows:

$$|\mathbf{x}_a^\top \hat{\boldsymbol{\theta}}_a - \mathbb{E}[r_a | \mathbf{x}_a]| \leq \alpha \sqrt{\mathbf{x}_a^\top \mathbf{A}_a^{-1} \mathbf{x}_a}, \text{ where } \alpha = 1 + \sqrt{\ln(2/\delta)/2}. \tag{4}$$

Thus, the first term of the right side of Eq. (3) estimates the reward of arm a, and the second term works as the confidence interval of the average reward. Therefore, LinUCB calculates the upper confidence bound of the reward of each arm, similar to UCB algorithms. Algorithm 1 describes the LinUCB algorithm, which updates $\hat{\boldsymbol{\theta}}_a$ via supplementary matrix $\mathbf{A}$ and vector $\mathbf{b}$, at each time step.

Note that we introduce only the basic LinUCB framework for simplicity of the paper. The authors of LinUCB also presented an extended framework in which a feature vector models both a visiting user and an article available at time t [5, 15].

3   LinUCT and Variants

In the original LinUCB, it is assumed that each arm a has its own coefficient vector $\boldsymbol{\theta}_a^*$; therefore matrix $\mathbf{A}_a$ and vector $\mathbf{b}_a$ are maintained individually. However, this configuration prevents LinUCB from generalizing information among positions when we model a move as an arm in deterministic games. In contrast, it is reasonable to assume that the expected rewards are under the control of a common $\boldsymbol{\theta}^*$, for all nodes in a searched game tree. Hereafter, we use a common $\boldsymbol{\theta}^*$, matrix $\mathbf{A}$, and vector $\mathbf{b}$ (without subscripts). This decision follows the observation that a common evaluation function is used throughout search by minimax-based methods.

Algorithm 2. Supplementary Procedures in LinUCT

1: **procedure** LINUCT-INITIALIZE
2:     $\mathbf{A} \leftarrow \mathbf{I}_{d \times d}$
3:     $\mathbf{b} \leftarrow \mathbf{0}_{1 \times d}$
4: **procedure** BACK-PROPAGATION-ADDED(path, Δ)
5:     **for** $s \in$ path **do**
6:         $\mathbf{A} \leftarrow \mathbf{A} + \mathbf{x}_s \mathbf{x}_s^\top$
7:         $\mathbf{b} \leftarrow \mathbf{b} + \Delta_s \mathbf{x}_s$

3.1 LinUCT$_{\text{PLAIN}}$: Basic LinUCT

Here we introduce LinUCT$_{\text{PLAIN}}$, which is a straightforward application of Lin-UCB to MCTS. In LinUCT$_{\text{PLAIN}}$, the selection step described in Sect. 2 employs LinUCB with the following adjustment in counting the number of simulations:

$$\text{LinUCB'}(s,a) = \mathbf{x}_a^\top \hat{\boldsymbol{\theta}} + \alpha \sqrt{\mathbf{x}_a^\top \cdot \frac{N_{s_0}}{N_a} \mathbf{A}^{-1} \cdot \mathbf{x}_a}, \tag{5}$$

where s_0 is the root node of a given search tree.

Algorithm 2 shows the supplementary procedures used in LinUCT$_{\text{PLAIN}}$. Procedure LINUCT-INITIALIZE initializes the global variables $\mathbf{A}$ and $\mathbf{b}$. After each simulation, in addition to the standard back-propagation process of N_s and $\overline{X}_s$ in MCTS, procedure BACK-PROPAGATION-ADDED updates variables $\mathbf{A}$ and $\mathbf{b}$ for each node s in the path from the root using the result of a playout Δ. Variable Δ_s is the relative reward of Δ with respect to the player of s. Consequently, matrix $\mathbf{A}$ is updated multiple times for each playout, while it is updated exactly once in the original LinUCB. This makes the second term for exploration in Eq. (3) too small too rapidly. Therefore, as in Eq. (5), we scale the elements in matrix $\mathbf{A}$ by the total number of playouts divided by the number of visits to the node.

3.2 LinUCT$_{\text{RAVE}}$: LinUCB Combined with UCB1 in RAVE Form

A concern with LinUCT$_{\text{PLAIN}}$ is that it evaluates a node only by its static feature vector. Consequently, the descendant node information is completely ignored, which is apparently problematic, because in traditional minimax search methods, the minimax value of as deep search as possible is preferable to a mere evaluation function's value for the root node.

LinUCT$_{\text{RAVE}}$, a combination of LinUCB and UCB1, resolves this problem by simply utilizing LinUCB as the RAVE heuristic function in Eq. (2).

$$\text{LinUCB}_{\text{RAVE}}(s,a) = \beta(s)\,\text{LinUCB'}(s,a) + (1 - \beta(s))\,\text{UCB1}(s,a) \tag{6}$$

The value converges to UCB1 value as the original RAVE presented in Eq. (2) does. In addition, LinUCT$_{\text{RAVE}}$ makes the idea of RAVE more domain-independent. While the original RAVE assumes that the value of a move is

---

Algorithm 3. Back-propagation process of $\text{LinUCT}_{\text{FP}}$

1: **procedure** BACK-PROPAGATION-ADDED(path, Δ)
2:    **for** $s \in$ path **do**
3:        $\mathbf{A} \leftarrow \mathbf{A} + \mathbf{x}_s \mathbf{x}_s^\top$
4:        $\mathbf{b} \leftarrow \mathbf{b} + \Delta_s \mathbf{x}_s$
5:        $p \leftarrow$ parent of s
6:        **if** p is not null **then**
7:            $\mathbf{x}_p \leftarrow (1 - \gamma)\,\mathbf{x}_p + \gamma\,\mathbf{x}_s$

---

independent of move order in most cases in a target game. This assumption holds in Go; however, apparently it does not hold in chess. In contrast to the original RAVE, $\text{LinUCT}_{\text{RAVE}}$ can be applied to any game where a positional feature vector is available.

3.3 $\text{LinUCT}_{\text{FP}}$: LinUCB with Propagation of Features

$\text{LinUCT}_{\text{FP}}$ (feature propagation) is a completely different solution that considers subtrees. In $\text{LinUCT}_{\text{FP}}$, by recursively propagating the feature vector of a node to that of its parent, the LinUCB value calculated using Eq. (3) reflects the expected rewards of playouts through the node. Algorithm 3 describes the modified back-propagation process used in $\text{LinUCT}_{\text{FP}}$, where $\gamma \in (0, 1)$ controls the learning rate of a feature vector. We also present $\text{LinUCT}_{\text{RAVE-FP}}$, which incorporates this propagation scheme into $\text{LinUCT}_{\text{RAVE}}$.

4   Incremental Random Game Tree with Feature Vectors

Here we introduce an extension to existing random game-tree models. Incremental random trees (or P-game) have served as domain-independent test sets for evaluation of various search algorithms [7,13,14,18,20]. In this context, a random value is assigned to each edge, and the game theoretical value of a leaf is defined as the summation of the edge values in the path from the root. Moreover, for an internal node, the same summation can serve as a heuristic score returned by an evaluation function for that node. The advantages of this model are that (1) the search space can be controlled easily via the width and height, and (2) a correlation between the heuristic score of a node and that of its descendants is produced, which is expected in real games. Here, we extend the trees such that each node has a feature vector while preserving the main property of incremental random trees.

In our new trees, each tree has its own hidden d-dimensional vector $\boldsymbol{\theta}^* \in \mathbb{R}^d$, which cannot be observed by search algorithms. In addition, each node in a tree has two d-dimensional binary feature vectors (the one is for a player to move, and the other for the opponent): $\mathbf{x}_m, \mathbf{x}_o \in \{0,1\}^d$. In MCTS, a leaf returns binary reward $r = \{0,1\}$ for each playout, and the expected reward $\mathbb{E}(r)$ is defined by these vectors as follows:

$$\mathbb{E}(r) = (\mathbf{x}_m - \mathbf{x}_o)^\top \boldsymbol{\theta}^* + 0.5. \tag{7}$$

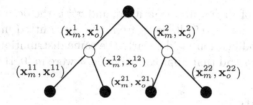

Fig. 1. Example of an incremental random tree with feature vectors.

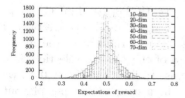

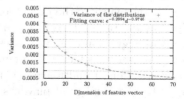

Fig. 2. Reward distributions   **Fig. 3.** Variance v.s. dimension

Similarly, the inner product $(\mathbf{x}_m - \mathbf{x}_o)^\top \hat{\boldsymbol{\theta}}$ gives the estimation of reward in LinUCB algorithms. This setting is equivalent to a Bernoulli multi-armed bandit problem when a tree depth is one.

To model a move in a two-player game, the feature vectors of each child $(\mathbf{x}'_m, \mathbf{x}'_o)$ are computed using those of the parent $(\mathbf{x}_m, \mathbf{x}_o)$. The feature vectors of a child inherit those of the parent with changing their owners, i.e., $\mathbf{x}'_m = \mathbf{x}_o$. This swap of vectors models the zero-sum property. Then, with a fixed probability p, each element (bit) is flipped. These flips correspond to changes in a board made by a move. Figure 1 shows an example of our tree. The left child of the root has feature vectors $(\mathbf{x}_m^1, \mathbf{x}_o^1)$, and its left child has feature vectors $(\mathbf{x}_m^{11}, \mathbf{x}_o^{11})$. These vectors are similar to $(\mathbf{x}_o^1, \mathbf{x}_m^1)$. For integer $i \in [0, d]$, the i-th element of $\mathbf{x}_m^{11}$ has the same or flipped value as that of the corresponding element of $\mathbf{x}_o^1$ with the probability p or $1 - p$, respectively.

Here we discuss some of the properties of our trees. We generated 100 trees by varying the dimension d of features. Note that the width and depth of a tree are fixed to 100 and 1, respectively. Figure 2 shows the histogram of the expected rewards, and Fig. 3 gives their variances with a fitted curve. As can be seen, the distributions with higher dimension have a smaller variance. The reason for this is that each element of $\boldsymbol{\theta}^*$ is randomly set in $[-0.5/d, 0.5/d]$, to ensure that the probability given in Eq. (7) is within $[0, 1]$. However, the difference in variances may cause a problem because it is difficult for UCB algorithms to select the best arm when the reward of the second best arm is close to that of the best one. Therefore, we adjust the rewards according to the observed variances, and use the following formula rather than Eq. (7):

$$\mathbb{E}(r) = \min(1.0, \max(0.0, \frac{(\mathbf{x}_m - \mathbf{x}_o)^\top \boldsymbol{\theta}}{\sqrt{e^{-3.24994}d^{-0.9746}/\sigma'^2}} + 0.5)), \tag{8}$$

where e is the base of the natural logarithm, and σ'^2 is the desired variance. The constants -3.24994 and -0.9746 come from the curve fitted in Fig. 3. We confirmed that this modification yields nearly the same distributions of rewards for various dimensions d and that approximately 96 % were in $[0, 1]$ for $\sigma' = 0.5/3$.

5 Experiments

We performed experiments with various configurations of artificial trees to evaluate the performance of the proposed algorithms ($\text{LinUCT}_{\text{PLAIN}}$, $\text{LinUCT}_{\text{RAVE}}$, $\text{LinUCT}_{\text{FP}}$, and $\text{LinUCT}_{\text{RAVE-FP}}$) and UCT. These algorithms were investigated in terms of regrets and failure rates following the experiments in the literature [13]. *Regret R*, which is the cumulative loss of the player's selection, is a standard criterion in multi-armed bandit problems. To fit in the range $[0, 1]$, we divide R by the number of playouts n and use the average regret per playout $R/n = \mu^* - 1/n \sum_{t=1}^{n} \mu_{i_t}$. Here, μ^* is the expectation of reward of the optimal (maximum) move, and μ_{i_t} is the expected reward of the pulled arm i_t at time t. For game trees, we defined μ_i for each child i of the root as the theoretical minimax value of node i (i.e., the expected reward of the leaf of the principal variation). The *failure rate* is the rate by which the algorithm fails to choose the optimal move at the root. For each tree instance, each algorithm and each time t, whether the algorithm fails is determined by whether the most visited move in the root so far is the optimal move. By averaging them over tree instances, we obtain the failure rate of each algorithm at time t.

Instances of trees were generated randomly as described in Sect. 4. The parameter p for controlling the similarity between a parent and a child was fixed to 0.1, while the dimension d of feature vectors was selected according the number of leaves. In addition, we removed trees in which the optimal move is not unique. Each MCTS algorithm grows its search tree iteratively until it covers all nodes of the generated tree. All MCTS algorithms expand the children of a node at the second visit to the node unless the node is a leaf of the generated tree. For each playout, a move is selected randomly until it reaches the leaf. Then, the reward is set randomly by the probability associated with the node given in Eq. (8).

5.1 Robustness with Respect to Parameters

LinUCT algorithms depend on the parameters, i.e., exploration parameter α in the LinUCB value in Eq. (3), k in $\text{LinUCB}_{\text{RAVE}}$ and its variants, and propagating rate γ for $\text{LinUCT}_{\text{FP}}$. To observe the dependency of performance of LinUCT algorithms on these parameters, we evaluated the combinations of various configurations: for α, the values $1.0, 1.59, 1.83$, and 2.22 were tested, where 1.0 is the minimum value and the rest correspond to $\delta = 1.0, 0.5$, and 0.1 in Eq. (4), respectively. For k and γ, the values $100, 1000$, and 10000 and $0.1, 0.01$, and 0.001 were tested, respectively. To observe the dependence on the tree size, various pairs of (depth, branching factor) were tested: $(1, 256), (2, 16), (4, 4)$ and $(8, 2)$. Note

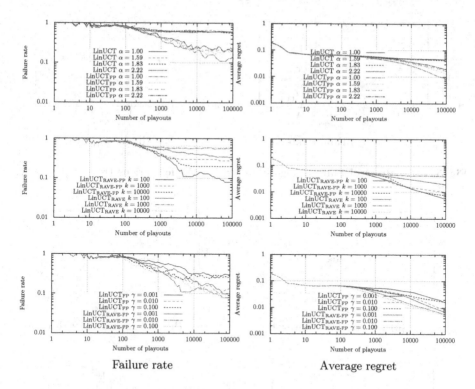

Fig. 4. Effects by constants α, k, and γ: depth = 4, width = 4, $d = 8$

that the dimension of feature vectors d was set to 8. Therefore each tree has exactly 256 leaves; thus, LinUCB can distinguish leaves in ideal cases.

The top, middle, and bottom two panels in Fig. 4 show the failure rates and average regrets for each algorithm with varying α, k, and γ, respectively. Each point represents the average over 100 trees. As can be seen, the constants $\alpha = 1.0$, $k = 100$, and $\gamma = 0.01$ are slightly better than others, although the differences among the algorithms are more crucial than those between the parameters for the same algorithm. Therefore, we used these parameters for the rest of our experiments. Note that we only show the results for trees $(4, 4)$, because, the results obtained with different tree configurations were similar.

5.2   Comparison with UCT

We compared LinUCT algorithms to UCT, which is the standard algorithm in MCTS. For this experiment, the depth of the tree was fixed to four, while various branching factors (4–16) were used. Figure 5 shows the failure rates and regrets for each algorithm, where each point (x, y) is the branching factor of trees for x, and the average failure rate or regret over 100 trees at time 10,000 for each algorithm for y. As expected, LinUCT$_{\text{PLAIN}}$ resulted in the highest failure rate (i.e., worst performance) for most cases. UCT outperformed the others for very

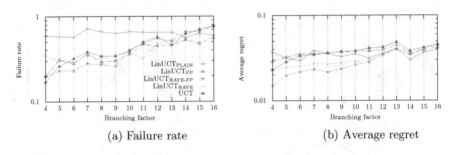

(a) Failure rate   (b) Average regret

Fig. 5. Comparison of LinUCT algorithms with UCT for various trees: depth = 4, $d = 16$

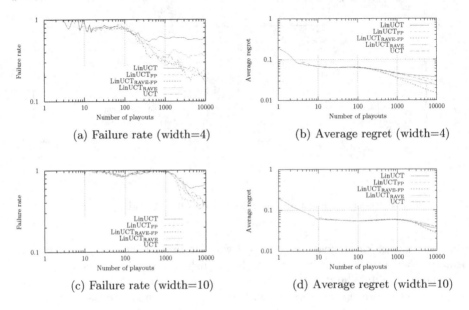

(a) Failure rate (width=4)   (b) Average regret (width=4)

(c) Failure rate (width=10)   (d) Average regret (width=10)

Fig. 6. Performances of each algorithm (depth = 4, $d = 16$)

small trees (e.g., branching factor four), LinUCT$_{\text{RAVE-FP}}$ performed better with a branching factor of 5–9, and LinUCT$_{\text{FP}}$ achieved good results for trees with a branching factor of greater than 12. These results suggest that LinUCT$_{\text{FP}}$ is effective in games that have a relatively large branching factor. Figure 6 shows that the failure rates and average regrets decreased along with an increased number of playouts. The performance of compared algorithms is similar up to a certain point; however, they differ substantially at time 10,000.

6 Conclusion

We presented a family of LinUCT algorithms that incorporate LinUCB into UCT for tree search: LinUCT$_{\text{PLAIN}}$, LinUCT$_{\text{RAVE}}$, LinUCT$_{\text{FP}}$, and LinUCT$_{\text{RAVE-FP}}$.

LinUCT$_{PLAIN}$ is the simplest algorithm in which the LinUCB value is used rather than the UCB1 value. However, there is room for improvement. Feature vectors observed for a node by the algorithm do not contain the information about the structure of the subtree expanded thus far. To address the problem, we incorporated existing techniques: a RAVE framework and feature propagation. LinUCT$_{RAVE}$ combines LinUCB and UCB1 in a RAVE framework. LinUCT$_{FP}$ is a modified version of LinUCT$_{PLAIN}$ in which the feature vectors of descendants are propagated to ancestors. LinUCT$_{RAVE\text{-}FP}$ is a combination of LinUCT$_{RAVE}$ and LinUCT$_{FP}$.

Experiments were performed with incremental random trees to assess the proposed algorithms in terms of the failure rates and regrets. In these experiments, each random tree was extended to have its own coefficient vector and feature vectors for each node, where the expected reward at each leaf is defined by the inner product of the feature vector and the coefficient vector. The results obtained with trees of a branching factor of 4–16 showed that LinUCT$_{RAVE}$, LinUCT$_{FP}$ and LinUCT$_{RAVE\text{-}FP}$ outperformed UCT, with the exception of small trees, and LinUCT$_{FP}$ demonstrated the best performance with a branching factor greater than 11.

There are two directions for future work. The most important direction for future work would be to examine the practical performance of the proposed family of LinUCT algorithm with major games such as Go. The other interesting direction is convergence analysis of the proposed LinUCT algorithms.

Acknowledgement. A part of this work was supported by JSPS KAKENHI Grant Number 25330432.

References

1. Auer, P., Cesa-Bianchi, N., Fischer, P.: Finite-time analysis of the multiarmed bandit problem. Mach. Learn. **47**(2–3), 235–256 (2002)
2. Bouzy, B., Cazenave, T.: Computer Go: an AI-oriented survey. Artif. Intell. **132**(1), 39–103 (2001)
3. Browne, C., Powley, E., Whitehouse, D., Lucas, S., Cowling, P., Rohlfshagen, P., Tavener, S., Perez, D., Samothrakis, S., Colton, S.: A survey of monte carlo tree search methods. IEEE Trans. Comput. Intell. AI Games **4**(1), 1–43 (2012)
4. Bubeck, S., Cesa-Bianchi, N.: Regret analysis of stochastic and nonstochastic multi-armed bandit problems. Found. Trends Mach. Learn. **5**(1), 1–122 (2012)
5. Chu, W., Li, L., Reyzin, L., Schapire, R.E.: Contextual bandits with linear pay-off functions. In: Gordon, G.J., Dunson, D.B., Dudík, M. (eds.) Proceedings of the Fourteenth International Conference on Artificial Intelligence and Statistics, AISTATS 2011. JMLR Proceedings, vol. 15, pp. 208–214. JMLR.org (2011)
6. Coulom, R.: Computing elo ratings of move patterns in the game of go. ICGA J. **30**(4), 198–208 (2007)
7. Furtak, T., Buro, M.: Minimum proof graphs and fastest-cut-first search heuristics. In: Boutilier, C. (ed.) Proceedings of the 21st IJCAI, pp. 492–498 (2009)

8. Gelly, S., Kocsis, L., Schoenauer, M., Sebag, M., Silver, D., Szepesvári, C., Teytaud, O.: The grand challenge of computer go: Monte carlo tree search and extensions. Commun. ACM **55**(3), 106–113 (2012)

9. Gelly, S., Silver, D.: Combining online and offline knowledge in uct. In: Proceedings of the 24th ICML, pp. 273–280. ACM (2007)

10. Gelly, S., Silver, D.: Monte-carlo tree search and rapid action value estimation in computer go. Artif. Intell. **175**(11), 1856–1875 (2011)

11. Hoki, K., Kaneko, T.: Large-scale optimization for evaluation functions with minimax search. J. Artif. Intell. Res. **49**, 527–568 (2014)

12. Knuth, D.E., Moore, R.W.: An analysis of alpha-beta pruning. Artif. Intell. **6**(4), 293–326 (1975)

13. Kocsis, L., Szepesvári, C.: Bandit based Monte-Carlo planning. In: Fürnkranz, J., Scheffer, T., Spiliopoulou, M. (eds.) ECML 2006. LNCS (LNAI), vol. 4212, pp. 282–293. Springer, Heidelberg (2006)

14. Korf, R.E., Chickering, D.M.: Best-first minimax search. Artif. Intell. **84**, 299–337 (1996)

15. Li, L., Chu, W., Langford, J., Schapire, R.E.: A contextual-bandit approach to personalized news article recommendation. In: Proceedings of the 19th International Conference on World Wide Web, pp. 661–670. ACM (2010)

16. Rosin, C.: Multi-armed bandits with episode context. Ann. Math. Artif. Intell. **61**(3), 203–230 (2011). doi:10.1007/s10472-011-9258-6

17. Silver, D., Tesauro, G.: Monte-carlo simulation balancing. In: Proceedings of the 26th Annual ICML, pp. 945–952. ACM (2009)

18. Smith, S.J., Nau, D.S.: An analysis of forward pruning. In: AAAI, pp. 1386–1391 (1994)

19. Walsh, T.J., Szita, I., Diuk, C., Littman, M.L.: Exploring compact reinforcement-learning representations with linear regression. In: Proceedings of the Twenty-Fifth Conference on Uncertainty in Artificial Intelligence, pp. 591–598. AUAI Press (2009)

20. Yoshizoe, K., Kishimoto, A., Kaneko, T., Yoshimoto, H., Ishikawa, Y.: Scalable distributed monte-carlo tree search. In: the 4th SoCS, pp. 180–187 (2011)

Adapting Improved Upper Confidence Bounds for Monte-Carlo Tree Search

Yun-Ching Liu[(✉)] and Yoshimasa Tsuruoka

Department of Electrical Engineering and Information Systems,
University of Tokyo, Tokyo, Japan
cipherman@gmail.com

Abstract. The UCT algorithm, which combines the UCB algorithm and Monte-Carlo Tree Search (MCTS), is currently the most widely used variant of MCTS. Recently, a number of investigations into applying other bandit algorithms to MCTS have produced interesting results. In this research, we will investigate the possibility of combining the improved UCB algorithm, proposed by Auer et al. [2], with MCTS. However, various characteristics and properties of the improved UCB algorithm may not be ideal for a direct application to MCTS. Therefore, some modifications were made to the improved UCB algorithm, making it more suitable for the task of game-tree search. The Mi-UCT algorithm is the application of the modified UCB algorithm applied to trees. The performance of Mi-UCT is demonstrated on the games of 9×9 Go and 9×9 NoGo, and has shown to outperform the plain UCT algorithm when only a small number of playouts are given, and rougly on the same level when more playouts are available.

1 Introduction

The development of Monte-Carlo Tree Search (MCTS) has made significant impact on various fields of computer game play, especially the field of computer Go [6]. The UCT algorithm [3] is an MCTS algorithm that combines the UCB algorithm [4] and MCTS, by treating each node as a single instance of the multi-armed bandit problem. The UCT algorithm is one of the most prominent variants of the Monte-Carlo Tree Search [6].

Recently, various investigations have been carried out on exploring the possibility of applying other bandit algorithms to MCTS. The application of simple regret minimizing bandit algorithms has shown the potential to overcome some weaknesses of the UCT algorithm [7]. The sequential halving on trees (SHOT) [8] applies the sequential halving algorithm [11] to MCTS. The SHOT algorithm has various advantages over the UCT algorithm, and has demonstrated better performance on the game of NoGo. The H-MCTS algorithm [9] performs selection by the SHOT algorithm for nodes that are near to the root and the UCT algorithm for deeper nodes. H-MCTS has also shown superiority over the UCT in games such as 8×8 Amazons and 8×8 AtariGo. Applications of the KL-UCB [12] and Thompson sampling [13] to MCTS have also been investigated and produced some interesting results [10].

© Springer International Publishing Switzerland 2015
A. Plaat et al. (Eds.): ACG 2015, LNCS 9525, pp. 53–64, 2015.
DOI: 10.1007/978-3-319-27992-3_6

Algorithm 1. The Improved UCB Algorithm [2]

Input: A set of arms A, total number of trials T
Initialization: Expected regret $\Delta_0 \leftarrow 1$, a set of candidates arms $B_0 \leftarrow A$
for rounds $m = 0, 1, \cdots, \lfloor \frac{1}{2} \log_2 \frac{T}{e} \rfloor$ **do**

    (1) Arm Selection:
    for all arms $a_i \in B_m$ **do**
        for $n_m = \lceil \frac{2 \log(T \Delta_m^2)}{\Delta^2} \rceil$ times **do**
            sample the arm a_i and update its average reward w_i
        end for
    end for

    (2) Arm Elimination:
    $a_{max} \leftarrow$ MAXIMUMREWARDARM(B_m)
    for all arms $a_i \in B_m$ **do**
        if $(w_i + \sqrt{\frac{\log(T \Delta^2)}{2 n_m}}) < (w_{max} - \sqrt{\frac{\log(T \Delta^2)}{2 n_m}})$ **then**
            remove a_i from B_m
        end if
    end for

    (3) Update Δ_m
    $\Delta_{m+1} = \frac{\Delta_m}{2}$
end for

The improved UCB algorithm [2] is a modification of the UCB algorithm, and it has been shown that the improved UCB algorithm has a tighter regret upper bound than the UCB algorithm. In this research, we will explore the possibility of applying the improved UCB algorithm to MCTS. However, some (we mention early explorations and not an anytime algorithm) characteristics of the improved UCB algorithm may not be desirable for a direct application to MCTS. Therefore, we have made appropriate modifications to the improved UCB algorithm, making it more suitable for the task of game-tree search. We will demonstrate the impact and implications of the modifications we have made on the improved UCB algorithm in an empirical study under the conventional multi-armed bandit problem setting. We will introduce the *Mi-UCT* algorithm, which is the application of the modified improved UCB algorithm to MCTS. We will demonstrate the performance of the Mi-UCB algorithm on the game of 9×9 Go and 9×9 NoGo, which has shown to outperform the plain UCT when given a small number of playouts, and roughly on the same level when more playouts are given.

2   Applying Modified Improved UCB Algorithm to Trees

In this section we will first introduce the improved UCB algorithm. We will then proceed to make three main modifications to the improved UCB algorithm, and finally show how to apply the modified algorithm to Monte-Carlo Tree Search.

2.1   Improved UCB Algorithm

In the multi-armed bandit problem (MAB), a player is faced with a K-armed bandit, and the player can decide to pull one of the arms at each play. The bandit will produce a reward $r \in [0, 1]$ according to the arm that has been pulled. The distribution of the reward of each arm is unknown to the player. The objective of the player is to maximize the total amount of reward over T plays. Bandit algorithms are policies that the player can follow to achieve this goal. Equivalent to maximizing the total expected reward, bandit algorithms aim to minimize the cumulative regret, which is defined as

$$R_t = \sum_{t=1}^{T} r^* - r_{I_t},$$

where r^* is the expected mean reward of the optimal arm, and r_{I_t} is the received reward when the player chooses to play arm $I_t \in K$ at play $t \in T$. If a bandit algorithm can restrict the cumulative regret to the order of $O(\log T)$, it is said to be optimal [1]. The UCB algorithm [4], which is used in the UCT algorithm [3], is an optimal algorithm which restricts the cumulative regret to $O(\frac{K \log(T)}{\Delta})$, where Δ is the difference of expected reward between a suboptimal arm and the optimal arm. The improved UCB algorithm [2] is a modification of the UCB algorithm, and it can further restrict the growth of the cumulative regret to the order of $O(\frac{K \log(T\Delta^2)}{\Delta})$.

The improved UCB algorithm, shown in Algorithm 1, essentially maintains a candidate set B_m of potential optimal arms, and then proceeds to systematically eliminate arms which are estimated to be suboptimal from that set. A predetermined number of total plays T is given to the algorithm, and the plays are further divided into $\lfloor \frac{1}{2} \log_2(\frac{T}{e}) \rfloor$ rounds. Each round consists of three major steps. In the first step, the algorithm samples each arm that is in the candidate set $n_m = \lceil \frac{2 \log(T\Delta_m^2)}{\Delta_m^2} \rceil$ times. Next, the algorithm proceeds to remove the arms whose upper bounds of estimated expected reward are less than the lower bound of the current best arm. The estimated difference Δ_m is then halved in the final step. After each round, the expected reward of the arm a_i is effectively estimated as

$$w_i \pm \sqrt{\frac{\log(T\Delta_m^2)}{2n_m}} = w_i \pm \sqrt{\frac{\log(T\Delta_m^2) \cdot \Delta_m^2}{4 \log(T\Delta_m^2)}} = w_i \pm \frac{\Delta_m}{2},$$

where w_i is the current average reward received from arm a_i.

In the case when the total number of plays T is not predetermined, the improved UCB algorithm can be run in an episodic manner; a total of $T_0 = 2$ plays is given to algorithm in the initial episode, and the number of plays of subsequent episodes is given by $T_{\ell+1} = T_\ell^2$.

2.2   Modification of the Improved UCB Algorithm

Various characteristics of the improved UCB algorithm might be problematic for its application to MCTS. We mention two of the **characteristic modifications**.

Algorithm 2. Modified Improved UCB Algorithm

Input: A set of arms A, total number of trials T

Initialization: Expected regret $\Delta_0 \leftarrow 1$, arm count $N_m \leftarrow |A|$, plays till Δ_k update $T_{\Delta_0} \leftarrow n_0 \cdot N_m$, where $n_0 \leftarrow \lceil \frac{2\log(T\Delta_0^2)}{\Delta_0^2} \rceil$, number of times arm $a_i \in A$ has been sampled $t_i \leftarrow 0$.

for rounds $m = 0, 1, \cdots T$ **do**

   (1) **Sample Best Arm:**

   $a_{max} \leftarrow \arg\max_{i \in |A|}(w_i + \sqrt{\frac{\log(T\Delta_k^2) \cdot r_i}{2n_k}})$, where $r_i = \frac{T}{t_i}$

   $w_{max} \leftarrow \text{UPDATEMAXWINRATE}(A)$

   $t_i \leftarrow t_i + 1$

   x

   (2) **Arm Count Update:**

   for all arms a_i **do**

      if $(w_i + \sqrt{\frac{\log(T\Delta_k^2)}{2n_k}}) < (w_{max} - \sqrt{\frac{\log(T\Delta_k^2)}{2n_k}})$ **then**

         $N_m \leftarrow N_m - 1$

      end if

   end for

   (3) **Update Δ_k when Deadline T_{Δ_k} is Reached**

   if $m \geq T_{\Delta_k}$ **then**

      $\Delta_{k+1} = \frac{\Delta_k}{2}$

      $n_{k+1} \leftarrow \lceil \frac{2\log(T\Delta_{k+1}^2)}{\Delta_{k+1}^2} \rceil$

      $T_{\Delta_{k+1}} \leftarrow m + (n_{k+1} \cdot N_m)$

      $k \leftarrow k + 1$

   end if

end for

- **Early Explorations.** The improved UCB algorithm tries to find the optimal arm by *the process of elimination*. Therefore, in order to eliminate suboptimal arms as early as possible, it has the tendency to devote more plays to suboptimal arms in the early stages. This might not be ideal when it comes to MCTS, especially in situations when time and resources are rather restricted, because it may end up spending most of the time exploring irrelevant parts of the game tree, rather than searching deeper into more promising subtrees.
- **Not an Anytime Algorithm.** We note that (1) the improved UCB algorithm requires the total number of plays to be specified in advance, and (2) its major properties or theoretical guarantees may not hold if the algorithm is stopped prematurely. Since we are considering each node as a single instance of the MAB problem in MCTS, internal nodes which are deeper in the tree are most likely the instances that are prematurely stopped. So, on the one hand the "temporal" solutions provided by these nodes might be erroneous, and the effect of these errors may be magnified as they propagate upward to

the root node. On the other hand, it would be rather expensive to ensure that the required conditions are met for the improved UCB algorithms on each node, because the necessary amount of playouts will grow exponentially as the number of expanded node increases.

Therefore, we have the relevant some adjustments to the improved UCB algorithm before applying it to MCTS.

The modified improved UCB bandit algorithm is shown in Algorithm 2. The modifications try to retain the major characteristics of the improved UCB algorithm, especially the way the confidence bounds are updated and maintained. Nonetheless, we should note that these modifications will change the algorithm's behaviour, and the theoretical guarantees of the original algorithm may no longer be applicable.

Algorithmic Modifications. We have made two major adjustments to the algorithmic aspect of the improved UCB algorithm.

1. **Greedy Optimistic Sampling**. We only sample the arm that currently has the highest upper bound, rather than sampling every possible arm n_m times.
2. **Maintain Candidate Arm Count**. We will only maintain the count of potential optimal arms, instead of maintaining a candidate set.

Since we are only sampling the current best arm, we are effectively performing a more aggressive arm elimination; arms that are perceived to be suboptimal are not being sampled. Therefore, there is no longer a need for maintaining a candidate set.

However, the confidence bound in the improved UCB algorithm for arm a_i is defined as $w_i \pm \sqrt{\frac{\log(T\Delta_m^2)}{2n_m}}$, and the updates of Δ_m and n_m are both dictated by the number of plays in each round, which is determined by $(|B_m| \cdot n_m)$, i.e., the total number of plays that is needed to sample each arm in the candidate set B_m for n_m times. Therefore, in order to update the confidence bound we will need to maintain the count of potential optimal arms.

The implication of sampling the current best arm is that the guarantee for the estimated bound $w_i \pm \Delta_m$ to hold will be higher than the improved UCB algorithm, because the current best will likely be sampled more or equal to n_m times. This is desirable in game-tree search, since it would be more efficient to verify a variation is indeed the principal variation, than trying to identify and verify others are suboptimal.

Confidence Bound Modification. Since we have modified the algorithm to sample only the current best arm, the confidence bound for the current best arm should be tighter than other arms. Hence, an adjustment to the confidence bound is also needed.

In order to reflect the fact that the current best arm is sampled more than other arms, we have modified the definition of the confidence bound for arm a_i to

Algorithm 3. Modified Improved UCB Algorithm applied to Trees (Mi-UCT)

---

function MI-UCT(Node N)

   $best_{ucb} \leftarrow -\infty$

   for all child nodes n_i of N **do**

      if $n_i.t = 0$ **then**

         $n_i.ucb \leftarrow \infty$

      else

         $r_i \leftarrow N.episodeUpdate/n_i.t$

         $n_i.ucb \leftarrow n.w + \sqrt{\frac{\log(N.T \times N.\Delta^2) \times r_i}{2N.k}}$

      end if

      if $best_{ucb} \leq n_i.ucb$ **then**

         $best_{ucb} \leftarrow n_i.ucb$

         $n_{best} \leftarrow n_i$

      end if

   end for

   if $n_{best}.times = 0$ **then**

      $result \leftarrow$ RANDOMSIMULATION((n_{best}))

   else

      if n_{best} is not yet expanded **then** NODEEXPANSION((n_{best}))

      $result \leftarrow$ MI-UCT((n_{best}))

   end if

   $N.w \leftarrow (N.w \times N.t + result)/(N.t + 1)$

   $N.t \leftarrow N.t + 1$

   if $N.t \geq N.T$ **then**

      $N.\Delta \leftarrow 1$

      $N.T \leftarrow N.t + N.T \times N.T$

      $N.armCount \leftarrow$ Total number of child nodes

      $N.k \leftarrow \lceil \frac{2\log(N.T \times N.\Delta^2)}{N.\Delta^2} \rceil$

      $N.deltaUpdate \leftarrow N.t + N.k \times N.armCount$

   end if

   if $N.t \geq N.deltaUpdate$ **then**

      for all child nodes n_i of N **do**

         if $(n_i.w + \sqrt{\frac{\log(N.T \times N.\Delta^2)}{2n.k}}) < (N.w - \sqrt{\frac{\log(N.T \times N.\Delta^2)}{2n.k}})$ **then**

         $N.armCount \leftarrow N.armCount - 1$

         end if

      end for

      $N.\Delta \leftarrow \frac{N.\Delta}{2}$

      $N.k \leftarrow \lceil \frac{2\log(N.T \times N.\Delta^2)}{N.\Delta^2} \rceil$

      $N.deltaUpdate \leftarrow N.t + N.k \times N.armCount$

   end if

   return $result$

end function

function NODEEXPANSION(Node N)

   $N.\Delta \leftarrow 1$

   $N.T \leftarrow 2$

   $N.armCount \leftarrow$ Total number of child nodes

   $N.k \leftarrow \lceil \frac{2\log(N.t \times N.\Delta^2)}{N.\Delta^2} \rceil$

   $N.deltaUpdate \leftarrow N.k \times N.armCount$

end function

$$w_i \pm \sqrt{\frac{\log(T\Delta_m^2) \cdot r_i}{2n_m}},$$

where the factor $r_i = \frac{T}{t_i}$, and t_i is the number of times that the arm has been sampled. The more arm a_i is sampled, the smaller r_i will be, and hence the tighter is the confidence bound. Therefore, the expected reward of arm a_i will be estimated as

$$w_i \pm \sqrt{\frac{\log(T\Delta_m^2) \cdot r_i}{2n_m}} = w_i \pm \sqrt{\frac{\log(T\Delta_m^2) \cdot \Delta_m^2 \cdot r_i}{4\log(T\Delta_m^2)}} = w_i \pm \frac{\Delta_m}{2}\sqrt{r_i} = w_i \pm \frac{\Delta_m}{2}\sqrt{\frac{T}{t_i}}.$$

Since it would be more desirable that the total number of plays is not required in advance, we will run the modified improved UCB algorithm in an episodic fashion when we apply it to MCTS, i.e., assigning a total of $T_0 = 2$ plays to the algorithm in the initial episode, and $T_{\ell+1} = T_\ell^2$ plays in the subsequent episodes. After each episode, all the relevant terms in the confidence bound, such as Δ_m and n_m, will be re-initialized, and hence information from previous episodes will be lost. Therefore, in order to "share" information across episodes, we will not re-initialize r_i after each episode.

2.3   Modified Improved UCB Applied to Trees (Mi-UCT)

We will now introduce the application of the modified improved UCB algorithm to Monte-Carlo Tree Search, or the *Mi-UCT* algorithm. The details of the Mi-UCT algorithm are shown in Algorithm 3.

The Mi-UCT algorithm adopts the same game-tree expansion paradigm as the UCT algorithm, that is, the game tree is expanded over a number of iterations, and each iteration consists of four steps: *selection, expansion, simulation,* and *backpropagation* [3]. The difference is that the tree policy is replaced by the modified improved UCB algorithm. The modified improved UCB on each node is run in an episodic manner; a total of $T_0 = 2$ plays to the algorithm in the initial episode, and $T_{\ell+1} = T_\ell^2$ plays in the subsequent episodes.

The Mi-UCT algorithm keeps track of when $N.\Delta$ should be updated and the starting point of a new episode by using the variables $N.deltaUpdate$ and $N.T$, respectively. When the number of playouts $N.t$ of the node N reaches the updating deadline $N.deltaUpdate$, the algorithm halves the current estimated regret $N.\Delta$ and calculates the next deadline for halving $N.\Delta$. The variable $N.T$ marks the starting point of a new episode. Hence, when $N.t$ reaches $N.T$, the related variables $N.\Delta$ and $N.armCount$ are re-initialized, and the starting point $N.T$ of the next episode, along with the new $N.deltaUpdate$ are calculated.

3   Experimental Results

We will first examine how the various modifications we have made to the improved UCB algorithm affect its performance on the multi-armed bandit problem. Next, we will demonstrate the performance of the Mi-UCT algorithm against the plain UCT algorithm on the game of 9×9 Go and 9×9 NoGo.

3.1 Performance on Multi-armed Bandits Problem

The experimental settings follow the multi-armed bandit testbed that is specified in [5]. The results are averaged over 2000 randomly generated K-armed bandit tasks. We have set $K = 60$ to simulate more closely the conditions in which bandit algorithms will face when they are applied in MCTS for games that have a middle-high branching factor. The reward distribution of each bandit is a normal (Gaussian) distribution with the mean w_i, $i \in K$, and variance 1. The mean w_i of each bandit of every generated K-armed bandit task was randomly selected according to a normal distribution with mean 0 and variance 1.

The cumulative regret and optimal action percentage are shown in Figs. 1 and 2, respectively. The various results correspond to different algorithms as follows.

- **UCB**: the UCB algorithm.
- **I-UCB**: the improved UCB algorithm.
- **I-UCB (episodic)**: the improved UCB algorithm ran episodically.
- **Modified I-UCB (no r)**: only algorithmic modifications on the improved UCB algorithm.
- **Modified I-UCB (no r, episodic)**: only algorithmic modifications on the improved UCB algorithm ran episodically.
- **Modified I-UCB**: both algorithmic and confidence bound modifications on the improved UCB algorithm.
- **Modified I-UCB (episodic)**: both algorithmic and confidence bound modifications on the improved UCB algorithm ran episodically.

Contrary to theoretical analysis, we are surprised to observe the original improved UCB, both I-UCB and I-UCB (episodic), produced the worst cumulative regret. However, their optimal action percentages are increasing at a very rapid rate, and are likely to overtake the UCB algorithm if more plays are given. This suggests that the improved UCB algorithm does indeed devote more plays to exploration in the early stages.

The "slack" in the curves of the algorithms that were run episodically are the points when a new episode begins. Since the confidence bounds are essentially re-initialized after every episode, effectively extra explorations are performed. Therefore, there were extra penalties on the performance, and it can be clearly observed in the cumulative regret.

We can further see that by making only the algorithmic modification, to give Modified I-UCB (no r) and Modified I-UCB(no r, episodic), the optimal action percentage increases very rapidly, but it eventually plateaued and stuck to suboptimal arms. Their cumulative regret also increased linearly instead of logarithmically.

However, by adding the factor r_i to the confidence bound, the optimal action percentage increases rapidly and might even overtake the UCB algorithm if more plays are given. Although the optimal action percentage of the modified improved UCB, both Modified I-UCB and Modified I-UCB (episodic), are rapidly catching up with that of the UCB algorithm; there is still a significant gap between their cumulative regret.

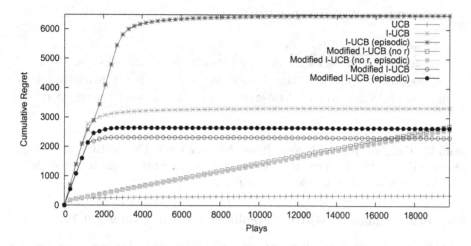

Fig. 1. Cumulative regret of various modifications on improved UCB algorithm

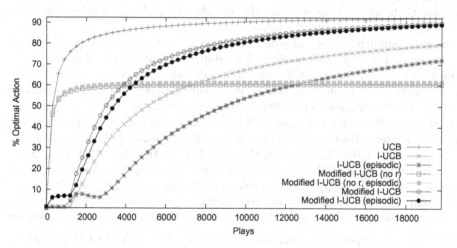

Fig. 2. Optimal arm percentage of various modifications on improved UCB algorithm

3.2 Performance of Mi-UCT Against Plain UCT on 9 × 9 Go

We will demonstrate the performance of the Mi-UCT algorithm against the plain UCT algorithm on the game of Go played on a 9 × 9 board.

For an effective comparison of the two algorithms, no performance enhancing heuristics were applied. The simulations are all pure random simulations without any patterns or simulation policies. A total of 1000 games were played for each constant C setting of the UCT algorithm, each taking turns to play Black. The total number of playouts was fixed to 1000, 3000, and 5000 for both algorithms.

The results are shown in Table 1. It can be observed that the performance of the Mi-UCT algorithm is quite stable against various constant C settings of the

Table 1. Win rate of Mi-UCT against plain UCT on 9 × 9 Go

Constant C	0.1	0.2	0.3	0.4	0.5	0.6	0.7	0.8	0.9
1000 playouts	57.1%	55.2%	57.5%	52.2%	58.6%	58.4%	55.8%	55.3%	54.5%
3000 playouts	50.8%	50.9%	50.3%	52.2%	52.2%	54.4%	56.5%	56.0%	54.1%
5000 playouts	54.3%	54.2%	52.4%	51.0%	52.4%	57.5%	54.9%	56.1%	55.3%

plain UCT algorithm, and is roughly on the same level. The Mi-UCT algorithm seems to have better performance when only 1000 playouts are given, but slightly deteriorates when more playouts are available.

3.3 Performance of Mi-UCT Against Plain UCT on 9 × 9 NoGo

We will demonstrate the performance of the Mi-UCT algorithm against the plain UCT algorithm on the game of NoGo played on a 9 × 9 board. NoGo is a misere version of the game of Go, in which the first player that has no legal moves other than capturing the opponent's stone loses.

All the simulations are all pure random simulations, and no extra heuristics or simulation policies were applied. A total of 1000 games were played for each constant C setting of the UCT algorithm, each taking turns to play Black. The total number of playouts was fixed to 1000, 3000, and 5000 for both algorithms.

The results are shown in Table 2. We can observe that the Mi-UCT algorithm significantly dominates the plain UCT algorithm when only 1000 playouts were given, and the performance deteriorates rapidly when more playouts are available, although it is still roughly on the same level as the plain UCT algorithm.

The results on both 9 × 9 Go and 9 × 9 NoGo suggest that the performance of the Mi-UCT algorithm is comparable to that of the plain UCT algorithm, but scalability seems poorer. Since the proposed modified improved UCB algorithm essentially estimates the expected reward of each bandit by $w_i + \frac{\Delta_m}{2}\sqrt{r_i}$, where $r_i = \sqrt{\frac{T}{t_i}}$, the exploration term converges slower than that of UCB algorithm, and hence more exploration might be needed for the modified improved UCB confidence bounds to converge to a "sufficiently good" estimate value; this might be the reason why Mi-UCT algorithm has poor scalability. Therefore, we might able to overcome this problem by trying other definitions for r_i.

Table 2. Win rate of Mi-UCT against plain UCT on 9 × 9 NoGo

Constant C	0.1	0.2	0.3	0.4	0.5	0.6	0.7	0.8	0.9
1000 playouts	58.5%	56.1%	61.4%	56.7%	57.4%	58.4%	59.6%	56.9%	57.8%
3000 playouts	50.3%	51.4%	53.1%	51.0%	49.6%	54.4%	56.0%	54.2%	53.9%
5000 playouts	45.8%	48.8%	48.5%	49.6%	55.1%	51.3%	51.3%	55.0%	52.7%

4 Conclusion

The improved UCB algorithm is a modification of the UCB algorithm, and has a better regret upper bound than the UCB algorithm. Various characteristics of the improved UCB algorithm, such as early exploration and not being an anytime algorithm, are not ideal for a direct application to MCTS. Therefore, we have made relevant modifications to the improved UCB algorithm, making it more suitable for the task of game-tree search. We have investigated the impact and implications of each modification through an empirical study under the conventional multi-armed bandit problem setting.

The Mi-UCT algorithm is the application of the modified improved UCB algorithm applied to Monte-Carlo Tree Search. We have demonstrated that it outperforms the plain UCT algorithm on both games of 9×9 Go and 9×9 NoGo when only a small number of playouts are given, and on comparable level with increased playouts. One possible way of improving the scalability would be trying other definitions of r_i in the modified improved UCB confidence bounds.

It would also be interesting to investigate the possibility of enhancing the performance of the Mi-UCT algorithm by combining it with commonly used heuristics [6] or develop new heuristics that are unique to the Mi-UCT algorithm. Finally, since the modifications made essentially changed the behaviour of the original algorithm, investigation into the theoretical properties of our modified improved UCB algorithm may provide further insight into the relation between bandit algorithms and Monte-Carlo Tree Search.

References

1. Lai, T.L., Robbins, H.: Asymptotically efficient adaptive allocation rules. Adv. Appl. Math. **6**(1), 4 (1985)
2. Auer, P., Ortner, R.: UCB revisited: improved regret bounds for the stochastic multi-armed bandit problem. Periodica Math. Hung. **61**, 1–2 (2010)
3. Kocsis, L., Szepesvári, C.: Bandit based monte-carlo planning. In: Fürnkranz, J., Scheffer, T., Spiliopoulou, M. (eds.) ECML 2006. LNCS (LNAI), vol. 4212, pp. 282–293. Springer, Heidelberg (2006)
4. Auer, P., Cesa-Bianchi, N., Fischer, P.: Finite-time analysis of the multiarmed bandit problem. Mach. Learn. **47**(2–3), 235–256 (2002)
5. Sutton, R.S., Barto, A.G.: Reinforcement Learning: An Introduction. MIT Press, Cambridge, MA (1998)
6. Browne, C.B., Powley, E., Whitehouse, D., Lucas, S.M., Cowling, P.I., Rohlfshagen, P., Tavener, S., Perez, D., Samothrakis, S., Colton, S.: A survey of monte carlo tree search methods. IEEE Trans. Comput. Intell. AI Games **4**(1), 1–43 (2012)
7. Tolpin, D., Shimony, S.E.: MCTS based on simple regret. In: Proceedings of the 26th AAAI Conference on Artificial Intelligence, pp. 570–576 (2012)
8. Cazenave, T.: Sequential halving applied to trees. IEEE Trans. Comput. Intell. AI Games **7**, 102–105 (2014)
9. Pepels, T., Cazenave, T., Winands, M.H.M., Lanctot, M.: Minimizing simple and cumulative regret in monte-carlo tree search. In: Cazenave, T., Winands, M.H.M., Björnsson, Y. (eds.) CGW 2014. CCIS, vol. 504, pp. 1–15. Springer, Heidelberg (2014)

10. Imagawa, T., Kaneko, T.: Applying multi armed bandit algorithms to MCTS and those analysis. In: Proceedings of the 19th Game Programming Workshop (GPW-14), pp. 145–150 (2014)
11. Karnin, Z., Koren, T., Oren, S.: Almost optimal exploration in multi-armed bandits. In: Proceedings of the 30th International Conference on Machine Learning (ICML'13), pp. 1238–1246 (2013)
12. Garivier, A., Cappe, A.: The KL-UCB algorithm for bounded stochastic bandits and beyond. In: Proceedings of 24th Annual Conference on Learning Theory (COLT '11), pp. 359–376 (2011)
13. Kaufmann, E., Korda, N., Munos, R.: Thompson sampling: an asymptotically optimal finite-time analysis. In: Bshouty, N.H., Stoltz, G., Vayatis, N., Zeugmann, T. (eds.) ALT 2012. LNCS, vol. 7568, pp. 199–213. Springer, Heidelberg (2012)

On Some Random Walk Games
with Diffusion Control

Ingo Althöfer[1]([✉]), Matthias Beckmann[1], and Friedrich Salzer[2]

[1] University of Jena, Jena, Germany
ingo.althoefer@uni-jena.de
[2] Frankfurt University, Frankfurt, Germany

Abstract. Random walks with discrete time steps and discrete state spaces have widely been studied for several decades. We investigate such walks as games with "Diffusion Control": a player (=controller) with certain intentions influences the random movements of the particle. In our models the controller decides only about the step size for a single particle. It turns out that this small amount of control is sufficient to cause the particle to stay in "premium regions" of the state space with surprisingly high probabilities.

1 Introduction

The paper presents results on discrete 1-player random walk games with diffusion control. In the games, time is running in discrete steps, and also the 1-dimensional state spaces are discrete (sometimes finite and sometimes infinite). A single particle is walking in the state space, and the player is an "outside" controller who has some influence on the movements of the particle. The controller is assumed to always have full information about the system. The control actions are small ones: in our games, the controller only decides whether the particle makes a small or a large fair random step in the current moment. Here, "fair" means that the particle makes steps to the left and to the right with the same probability $\frac{1}{2}$. Also the widths of the steps are the same for both directions.

The main results are: (i) a great deal of influence is possible and (ii) in finite (circular) state spaces the controller's influence is much larger compared to the controller's influence in infinite state spaces. Section 2 lists related research. In Sect. 3 a random walk on the integers is investigated. It allows to model a stock market situation: the player is a broker and tries to maximize the probability of getting a bonus. Her[1] optimal strategy is both simple and successful and tells why current bonus-systems in investment banking are problematic. Technically, our analysis is based on discrete dynamic programming and the Bellman equations. Section 4 investigates control of a random walk in a closed ring. In Sect. 5, we conclude with a discussion and the formulation of three open problems.

---

[1] For brevity, we use 'he' and 'his', whenever 'he or she' and 'his or her' are meant.

© Springer International Publishing Switzerland 2015
A. Plaat et al. (Eds.): ACG 2015, LNCS 9525, pp. 65–75, 2015.
DOI: 10.1007/978-3-319-27992-3_7

2   Related Work

Three different results are mentioned. First, the famous theorem by Polya makes clear why analysis of the control problem on the set of integers is non-trivial: certain events happen with probability 1, but have infinite expected run time. Second, we mention Ankirchner's recent analysis of a continuous diffusion control problem. Third, Beckmann's investigation of random walk models where the controller is allowed to induce drift, is discussed.

2.1   Polya's Recurrence Theorem on Fair Random Walks Without Control

Polya's theorem [12] reads as follows. A single particle is moving on the infinite set $\mathbb{Z}$ of integers in discrete time. It is going from i to $i - 1$ and $i + 1$ with probability $\frac{1}{2}$ each for all i and in all moments t. When the particle starts in 0, it will return to 0 with probability 1 earlier or later. However, the expected time until its first return is infinite. The same statement is true for dimension 2, but not for dimension 3 or higher.

2.2   Continuous Models with Diffusion Control

Recently, Ankirchner [4] proved a continuous counterpart of the casino model of Sect. 3: He assumed Brownian motion with 1-fold speed for all states $x \geq 0$ and double or triple speed in the negative part of the real axes. With this strategy, a particle starting in 0 stays in the positive numbers with probabilities $\frac{2}{3}$ and $\frac{3}{4}$, respectively. Ankirchner also found the old abstract McNamara paper [10] on optimal continuous diffusion control, where a bang-bang strategy is proven to be optimal for a whole class of reward functions (highest possible diffusion rate in the negative numbers, smallest possible in the positive numbers). Similar models have been investigated in ecology [14] and in insurance mathematics [5].

2.3   Beckmann's Evacuation Models

Even more recently, Beckmann [7] analysed random walk models where the controller is allowed to block one of the direct neighbour states of the particle. In a one-dimensional state space optimal control is trivial, in stale spaces of dimension 2 and all higher dimensions it becomes interesting. These "single direction blockades" allow the particle to travel with linear expected speed – in contrast to the square-root speed of particles in normal random walks and Brownian motion. In particular, single blockades make random walks recurrent for all finite dimensions, in contrast to the unblocked case (see [12]).

3   Play in (m, n)-Casinos

Below, we introduce a set of new models for actions of a broker at the stock market. The broker is working for her bank. By contract, at the end of the year she gets a bonus of fixed size when she has not made a loss in total in that year. Each day, she has to perform exactly one transaction (a small one or a big one; her choice), where the outcome is positive with probability $\frac{1}{2}$ and negative also with probability $\frac{1}{2}$. (So, the stock market is assumed to be a fair casino for each transaction.)

Each model has two integral parameters m and n, with $1 \leq m < n$. If the action is a small one, win and loss are $+m$ and $-m$, respectively. If the action is a big one, win and loss are $+n$ and $-n$, respectively. The broker knows her current balance and the remaining number of days. Then the day has arrived that the broker has to decide between a small action and a big action for that day. The broker wants to maximize the probability of getting a bonus. We call these models (m, n)-casinos.

For the two simplest cases $(1, 2)$ and $(1, 3)$ a complete theoretical analysis has been executed [2,11]. Due to space limitations we give only the recursions and plausibility arguments for both results. In Subsect. 3.3 the Theorem of Polya is used to make the $\frac{2}{3}$- and $\frac{3}{4}$-results plausible. Subsection 3.4 contains a conjecture based on computational runs for all pairs (m, n) with $1 \leq m < n \leq 10$.

3.1 The $(1, 2)$-Casino

We formalize this optimization problem by discrete dynamic programming [8], with the following variables and recursions. Let $p_t(i)$ be the probability for getting the bonus, when the broker "plays" optimally, has currently balance i, and there are still t days to be played. Then the starting values are

$$p_0(i) = 1 \; for \; all \; i \; \geq 0 \; and \; p_0(i) = 0 \; for \; all \; i < 0.$$

For all $t \geq 0$ the Bellman equations say

$$p_{t+1}(i) = \max\{0.5 \cdot p_t(i - 1) + 0.5 \cdot p_t(i + 1);$$
$$0.5 \cdot p_t(i - 2) + 0.5 \cdot p_t(i + 2)\}$$

for all i in $\mathbb{Z}$. The $p_t(i)$ can be computed with computer help, for instance, on a PC with moderate memory space for all $t < 1,024$ and all i between $-10,240$ and $+10,240$. Outcomes for the central state $i = 0$ are

$$p_1(0) = \frac{1}{2}, \qquad\qquad p_2(0) = \frac{3}{4},$$

$$p_3(0) = \frac{5}{8}, \qquad\qquad p_4(0) = \frac{11}{16},$$

$$p_5(0) = \frac{21}{32}, \qquad\qquad p_6(0) = \frac{43}{64} \ldots$$

In each step the denominator is doubled, and the nominator is doubled ± 1. For general t, this structure is proved in [11]. The proof is based on manipulations of modified equations for binomial coefficients.

The consequence is $\lim_{t \to \infty} p_t(0) = \frac{2}{3}$. In particular, for $t = 240$ (a realistic value for the number of trading days in a year) the bonus probability is already very close to $\frac{2}{3}$: the difference is smaller than $\frac{1}{2^{240}}$. A corresponding optimal strategy consists in small steps whenever $i \geq 0$ and big steps for all situations (i, t) with $i < 0$. One side observation: there are many situations (i, t) where both actions are equally optimal.

For people from investment banking, the optimal strategy does not come as a surprise. When a bonus depends only on the plus-minus sign of the final balance, the broker should act the more aggressive the more she is in the minus. If she would be allowed to make arbitrary non-negative stakes in each round, the optimal strategy would be the well known "doubling strategy": Start with amount 1 in round 1. If you win, stop immediately (by staking 0 in all successive rounds). If you lose, stake amount 2 in round 2. If you win, stop immediately. If you lose, put amount 4 in the third round etc.

3.2   The $(1, 3)$-Casino

A slightly different model assumes big actions to have step sizes $+3$ and -3 instead of $+2$ and -2, all other things remaining unchanged. Recursions analogous to those in Subsect. 3.1 can be set up. It turns out that $\lim p_t(0) = \frac{3}{4}$ for t to infinity. A corresponding optimal strategy makes small steps for all states $i \geq 0$ and large steps for all states $i < 0$, for all t [2]. The structure of the solution and the corresponding probabilities are simpler than those for the $(1, 2)$-casino, because all possible steps have odd size. So, a particle starting in state 0 at $t = 0$ will always be in the set of even numbers for all even values of t, and in the set of odd numbers for all odd values of t.

3.3   Making the $\frac{2}{3}$-Result for the $(1, 2)$-Casino Plausible with Help of Polya's Theorem

Computing "like in physics", the $\frac{2}{3}$-probability for the $(1, 2)$-casino becomes plausible as follows: A fair random walk on $\mathbb{Z}$ starting in state "1" needs in average infinitely ($=$ inf) many steps to reach state "0" for the first time. Analogously, a fair random walk starting in state "0" needs in average again inf many steps to reach state "-1" for the first time. So, walking from "$+1$" to "-1" takes on average $2 \cdot$ inf many steps. On the other hand, a random walk with $(+2, -2)$-steps and start in "-1" takes on average inf many steps to reach "$+1$" for the first time. So, in average one cycle from negative border "-1" to "$+1$" and back from "$+1$" to "-1" via "0" takes inf $+2 \cdot$ inf steps. The corresponding ratio $\frac{2 \cdot \text{inf}}{2 \cdot \text{inf} + \text{inf}} = \frac{2}{3}$ is exactly what holds for the $(1, 2)$-casino.

In analogy, in the $(1, 3)$-casino it takes inf $+$ inf $+$ inf $= 3 \cdot$ inf many steps to start in "$+2$" and reach "-1" via "$+1$" and "0", and only $1 \cdot$ inf steps to return from "-1" to "$+2$" with random steps of size 3. And $\frac{3 \cdot \text{inf}}{3 \cdot \text{inf} + \text{inf}} = \frac{3}{4}$.

3.4   More General (m,n)-Casinos

With computer help we have looked at all 31 cases (m, n) with $1 \leq m < n \leq 10$, where m and n have greatest common divisor 1. For the basic strategy (small steps in the non-negative numbers, large ones in the negative numbers) we computed $p_t(i)$ for all $t \leq 1,024$ and all i between $-10,240$ and $+10,240$. Like before, $p_t(0)$ is the probability to end in the non-negative numbers after t steps when the start is in 0 and optimal decisions are made. In the Appendix we give $p_{64}(0)$, $p_{256}(0)$, and $p_{1024}(0)$ for each pair (m, n). Also the expected limit values are given for each parameter pair (m, n). With that data and the plausibility argument of Subsect. 3.3 in mind, we formulate two conjectures.

Conjecture 3.1: In the $(1, n)$-casino, it is optimal to make random steps of size 1 for all states $i \geq 0$ and steps of size n for all states $i < 0$. As a consequence, a particle starting in 0 stays in the non-negative area asymptotically with probability $\frac{n}{(n+1)}$.

Conjecture 3.2: In the (m, n)-casino with $1 < m < n$, it is optimal to make random steps of size m for almost all states $i \geq 0$ and steps of size n for almost all states $i < 0$. As a result, a particle starting in 0 will stay in the non-negative area asymptotically with probability $\frac{n}{n+m}$.

4   The (m, n)-Casinos on a Circle

Now we leave the stock market model by changing the state space: from Z to a closed circle with $2p$ cells. Motivation for such spaces comes from physics and bio-physics: given some compact space, agility of a particle may depend on the region in the space. For instance, the group Braun at LMU (Ludwig-Maximilians-Universität München) investigates phenomena such as thermophoresis where temperature-dependent diffusion leads to movement of large molecules [13].

Figure 1 shows a circle with $2 \cdot p = 2 \cdot 13 = 26$ cells. The diffusion controller wants to steer the particle into $R = \{0, 1, \ldots, p-1\}$, the right half of the circle. For convenience, we name the complement of R by S; so $S = \{p, p+1, \ldots, 2p-1\}$.

4.1   The $(1, 2)$-Casino on the Circle

The recursions are similar to those in Subsect. 3.1. The only difference is that the state numbers are now meant modulo $2p$. Let $r_t(i)$ be the largest possible probability to end in R, when i is the current state of the particle and still t steps are to be walked. Then we have

$$r_0(i) = 1 \text{ for all } i \text{ in } R \text{ and } r_0(i) = 0 \text{ for all } i \text{ in } S.$$

For all $t \geq 0$ we obtain

$$r_{t+1}(i) = \max\{0.5 \cdot r_t(i-1) + 0.5 \cdot r_t(i+1);$$
$$0.5 \cdot r_t(i-2) + 0.5 \cdot r_t(i+2)\}$$

for all i. Modulo $2p$ means for instance that $r_t(-1) = r_t(2p-1)$.

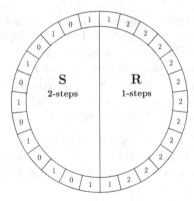

Fig. 1. The state space with $13 + 13$ cells. The right half, called R, is the preferred region. Also shown are the relative values of the stationary distribution for the $(1, 2)$-casino.

Observation: For all states i the $r_t(i)$ converge; for large p, $\lim_{t\to\infty} r_t(i) \approx \frac{4}{5}$. This limit can be achieved by the basic strategy with random steps of size 1 in R and steps of size 2 in S.

Fixing this basic strategy, the model can be viewed as a homogenous Markov process with unique stationary distribution $r(\ )$. Fortunately, this distribution has a simple structure. In the $(1, 2)$-model, for instance we achieve for $p = 13$

$$r(i) = \frac{2}{D} \text{ for all } i \in \{1, 2, \ldots, 11\}, r(0) = r(12) = \frac{1}{D},$$

$$r(i) = \frac{1}{D} \text{ for all } i \in \{13, 15, \ldots, 23, 25\}, \text{ and } r(i) = 0 \text{ for all } i \in \{14, 16, \ldots, 24\}.$$

The denominator D is to be chosen to obtain the total probability equal to 1. (So, for $p = 13$ we have $D = 33$.)

More important than the absolute value D is the observation that almost all states in R obtain 2-fold probability whereas only about half of the states in S obtain 1-fold probability (the other S-states become zero). Admittedly, it was on purpose that we took an odd number $p(= 13)$ for the example. Also for even numbers, state space R will end at almost $\frac{4}{5}$ of the probability. However, the structure of the stationary distribution $r(\ )$ is less simple.

4.2   The $(1, 3)$-Casino on the Circle

In the analogous model with steps of size 1 and 3 and state size $2p$, with $p = 3q + 1$ (for some integer q) the stationary distribution belonging to the basic strategy is $(1, 2, 3, 3, \ldots, 3, 2, 1; 1, 0, 0, 1, 0, 0, 1, \ldots, 1, 0, 0, 1)$. (The appropriate general denominator is omitted again.) The positions left of the semicolon belong to the states in R: most of them have 3-fold probability. The positions right of the semicolon stand for the states in S. Only every third position in S has a positive probability, namely a 1-fold. In total this means that R achieves almost $\frac{9}{10}$ of the total probability, when p is large.

4.3 On the General Circle Structure

Below, we give a heuristic explanation for the ratio $\frac{4}{5}$ in the (1,2)-casino, in analogy to the heuristic argument in Subsect. 3.3. Starting in state $2p - 1$ (or in p), the particle makes steps of size 2 until it reaches either $+1$ (or $p - 2$). In the average, these are about $1 \cdot \frac{p}{2}$ many steps. Starting in state $+1$ (or in $p - 2$), the particle makes steps of size 1, until it reaches either state $2p - 1$ or p. In the average, these are about $2 \cdot p$ steps. $2 \cdot p \approx 4 \cdot \frac{p}{2}$, hence the particle is in R for about $\frac{4}{4+1}$ of the time.

The term 4 contains one factor 2 from the double size of the steps in S; the other factor 2 results from the fact that the particle enters set R not in its border states 0 or $p - 1$, but in states 1 or $p - 2$. From there, it takes in the average about $2 \cdot (p-2) \approx 2p$ small steps to leave R again. (See Table 1 in the Appendix.)

Table 2 in the Appendix shows probability distributions between R and S for all pairs (m, n) with $1 \le m < n \le 10$, where m and n have greatest common divisor 1. In all cases we give the data for three circle sizes $p = 125, 250, 500$.

Observation 4.1: In the model with steps of size 1 in R and size n in S, the particle stays in R for about $\frac{n^2}{n^2+1}$ of the time. The corresponding stationary distribution has a simple structure for all p of the form $p = n \cdot q + 1$. For other forms of p, we have at least experimental evidence for the asymptotic ratio $\frac{n^2}{n^2+1}$.

Conjecture 4.2: In a large circle with steps of sizes m in R and n in S, the particle stays in R for about $\frac{n^2}{n^2+m^2}$ of the time. So, the parameters m and n enter the asymptotic distribution formula in squared form.

5 Discussion and Conclusions

Below, we provide a miscellaneous set of four subsections. They deal with (5.1) a Comparison, (5.2) Open Problems, (5.3) Other Bonus-/Malus-Functions, and (5.4) Micro Control.

5.1 Comparing $\mathbb{Z}$ and the Circle

In both state spaces it is optimal to make small steps in the "good" regions and large steps in the complement. However, the consequences are different. On the integral axes optimal control leads to ratio $\frac{n}{n+m}$, whereas on the circle the better ratio $\frac{n^2}{n^2+m^2}$ is achieved. So, the compact circle allows for much higher concentrations than the unbounded set $\mathbb{Z}$.

5.2 Open Problems

- We "need" formal proofs for $(1, n)$-casinos on $\mathbb{Z}$, for all $n \ge 4$.
- More generally, still missing are formal proofs for (m, n)-casino models for all $0 < m < n$, on $\mathbb{Z}$ and on the circle.

– In one of our models the parameters enter with exponent 1 (on $\mathbb{Z}$) and in the other model with exponent 2 (circle model). Do there exist reasonable $(1, n)$-models with exponent 3 or even higher for n? We think the answer is "no", but how to prove it?

5.3   Other Bonus-/Malus-Functions

$(1, 2)$-casino models with other bonus functions have completely different optimal strategies.

Example I: Variable bonus X, when the final balance is $X > 0$. Thus, there is no malus for $X \leq 0$. Then the optimal strategy is to always play big, i.e., random steps of size 2.

Example II: Variable malus $|Y|$, when the final balance is $Y < 0$. Thus, there is no bonus for $Y \geq 0$. Then the optimal strategy is to play always small, i.e. random steps of size 1.

The results for the examples bear consequences for the following more general theorem.

Theorem 5.1: Assume a bonus/malus function g on the set $\mathbb{Z}$ of integers. When g is convex on the whole range then the controller should always make big steps. For concave g it is optimal to make only small steps. The proof for the convexity case is a simple consequence of the convexity definition. From

$$g(x) < 0.5 \cdot g(x - 1) + 0.5 \cdot g(x + 1) \text{ for all } x$$

it follows immediately that also

$$0.5 \cdot g(x - 1) + 0.5 \cdot g(x + 1) < 0.5 \cdot g(x - 2) + 0.5 \cdot g(x + 2) \text{ for all } x.$$

An analogous statement is true for the case of concave functions g.

The function g in Example I is convex, the function g in Example II concave. The bonus function from Sect. 3 is neither convex nor concave. In general, functions with a mixture of convex and concave sections typically have non-trivial optimal control functions. Such mixed cases have been investigated in [10].

5.4   Remarks on Micro Control in Board Games

So far, micro control is not a hot topic in classical board games. However, sports such as sailing, gliding, paragliding, or car racing involve elements of diffusion control. Concerning MCTS bots for games such as Go, Hex or Havannah, human controllers with little influence may be able to improve the performance of the bots considerably. This happened, for instance, in the Havannah prize match [1], where one of the bot teams successfully modified very few Monte-Carlo parameters of the evaluation function between rounds. Very recently, small control actions helped to improve the performance of a human+computer team in Go on high-dan level [3]. An early investigation of Random Walk games can be found in [6].

Acknowledgments. Torsten Grotendiek [9] did rudimentary experimental analysis of the $(1, 2)$-model on the circle in his dissertation. Students from Ingo Althöfer's courses helped by active feedback on the casino models. Thanks go in particular to Bjoern Blenkers, Thomas Hetz, Manuel Lenhardt, and Maximilian Schwarz. Thanks will also go to two anonymous referees for their helpful criticism.

Appendix: Data for (m, n)-Casinos

Below, only the data for those pairs (m, n) are presented where m and n have greatest common divisor 1. In Table 1, for each tuple the probabilities for $t = 64, 256$, and 1024 are listed, as well as the suspected limit value $\frac{n}{n+m}$. In Table 2, for each tuple (m, n) the values of the stationary distributions for $p = 125, 250$ and 500 are listed as well as the suspected limit value $\frac{n^2}{n^2+m^2}$.

Table 1. (m, n)-casino on $\mathbb{Z}$

steps	(1,2)	(1,3)	(1,4)	(1,5)	(1,6)	(1,7)	(1,8)
64	0.666667	0.750000	0.779447	0.816419	0.827122	0.849332	0.868798
256	0.666667	0.750000	0.789957	0.824987	0.842735	0.862446	0.872004
1024	0.666667	0.750000	0.795005	0.829173	0.850000	0.868756	0.880547
lim	0.666667	0.750000	0.800000	0.833333	0.857143	0.875000	0.888889
steps	(1,9)	(1,10)	(2,3)	(2,5)	(2,7)	(2,9)	(3,4)
64	0.853216	0.869981	0.600000	0.698222	0.755097	0.786544	0.571609
256	0.884895	0.890578	0.600000	0.706992	0.766552	0.804225	0.571429
1024	0.892502	0.899981	0.600000	0.710703	0.772220	0.811342	0.571429
lim	0.900000	0.909091	0.600000	0.714286	0.777778	0.818182	0.571429
steps	(3,5)	(3,7)	(3,8)	(3,10)	(4,5)	(4,7)	(4,9)
64	0.625000	0.689351	0.703005	0.750332	0.557630	0.626214	0.673888
256	0.625000	0.694938	0.717800	0.757499	0.555556	0.631586	0.684221
1024	0.625000	0.697497	0.722695	0.763451	0.555556	0.634070	0.688430
lim	0.625000	0.700000	0.727273	0.769231	0.555556	0.636364	0.692308
steps	(5,6)	(5,7)	(5,8)	(5,9)	(6,7)	(7,8)	(7,9)
64	0.551902	0.583350	0.603999	0.634489	0.550146	0.549748	0.563288
256	0.545457	0.583333	0.611152	0.639180	0.538502	0.533566	0.562500
1024	0.545455	0.583333	0.613428	0.641062	0.538462	0.533333	0.562500
lim	0.545455	0.583333	0.615385	0.642857	0.538462	0.533333	0.562500
steps	(7,10)	(8,9)	(9,10)				
64	0.583356	0.549682	0.549674				
256	0.584860	0.530137	0.527881				
1024	0.586722	0.529412	0.526316				
lim	0.588235	0.529412	0.526316				

Table 2. (m, n)-casino on a circle with $2p$ cells

size p	(1.2)	(1.3)	(1.4)	(1.5)	(1.6)	(1.7)	(1.8)
125	0.797428	0.897106	0.938462	0.959119	0.970815	0.978054	0.982854
250	0.798722	0.898551	0.939839	0.960342	0.971909	0.979045	0.983747
500	0.799361	0.899279	0.940510	0.960944	0.972443	0.979526	0.984187
lim	0.800000	0.900000	0.941176	0.961538	0.972973	0.980000	0.984615
size p	(1,9)	(1,10)	(2,3)	(2,5)	(2,7)	(2,9)	(3,4)
125	0.986184	0.988612	0.688889	0.856354	0.918861	0.947761	0.636366
250	0.987016	0.989370	0.690608	0.859209	0.921714	0.950393	0.638177
500	0.987415	0.989740	0.691453	0.860641	0.923124	0.951676	0.639080
lim	0.987805	0.990099	0.692308	0.862069	0.924528	0.952941	0.640000
size p	(3,5)	(3,7)	(3,8)	(3,10)	(4,5)	(4,7)	(4,9)
125	0.729167	0.836413	0.867961	0.908666	0.606061	0.744969	0.824084
250	0.732193	0.840627	0.872340	0.913140	0.607859	0.749383	0.829533
500	0.733743	0.842716	0.874540	0.915297	0.608806	0.751617	0.832284
lim	0.735294	0.844828	0.876712	0.917431	0.609756	0.753846	0.835052
size p	(5,6)	(5,7)	(5,8)	(5,9)	(6,7)	(7,8)	(7,9)
125	0.586161	0.654957	0.709142	0.752725	0.572305	0.562445	0.615415
250	0.588234	0.658530	0.714284	0.758389	0.574490	0.564431	0.619221
500	0.589205	0.660377	0.716672	0.761249	0.575508	0.565399	0.621211
lim	0.590164	0.662162	0.719101	0.764151	0.576471	0.566372	0.623077
size p	(7,10)	(8,9)	(9,10)				
125	0.660562	0.554180	0.548682				
250	0.665900	0.556654	0.550442				
500	0.668524	0.557640	0.551476				
lim	0.671141	0.558621	0.552486				

References

1. Althöfer, I.: Havannah - the old man and the C's. ICGA J. **35**, 114–119 (2012)
2. Althöfer, I.: Theoretical analysis of the (1,3)-casino model. Working paper, Jena University, Faculty of Mathematics and Computer Science, September 2014
3. Althöfer, I., Marz, M., Kaitschick, S.: Computer-aided Go on high-dan level. Submitted for Presentation in the Workshop "Go and Science" at the European Go Congress 2015, Liberec (2015)
4. Ankirchner, S.: The answer is two third. Personal communication, July 2014
5. Asmussen, S., Taksarb, M.: Controlled diffusion models for optimal dividend payout. Insur. Math. Econ. **20**, 1–15 (1997)
6. Bachelier, L.: Le Jeu, la Chance et le Hasard (1914). (Republished in 2012 by Nabu Press)

7. Beckmann, M.: On some evacuation games with random walks. In: Plaat, A., van den Herik, J., Kosters, W. (eds.) Advances in Computer Games. LNCS, vol. 9525, pp. 89–99. Springer, Heidelberg (2015)
8. Bellman, R.: Dynamic Programming. Princeton University Press, Princeton (1957)
9. Grotendiek, T.: Analyse einiger Random Walk-Entscheidungsmodelle. Doctoral dissertation, Universitaet Bielefeld, Fakultaet fuer Mathematik (1996)
10. McNamara, J.M.: Optimal control of the diffusion coefficient of a simple diffusion process. Math. Oper. Res. **8**, 373–380 (1983)
11. Mitschunas, J.: Theoretical analysis of the (1,2)-casino model. Working paper, Jena University, Faculty of Mathematics and Computer Science, April 2015
12. Polya, G.: Über eine aufgabe betreffend die irrfahrt im strassennetz. Math. Ann. **84**, 149–160 (1921)
13. Reichl, M.R., Braun, D.: Thermophoretic manipulation of molecules inside living cells. J. Am. Chem. Soc. **136**, 15955–15960 (2014)
14. Stephens, D.W.: The logic of risk sensitive foraging processes. Anim. Behav. **29**, 628–629 (1981)

Go Complexities

Abdallah Saffidine[1]([⊠]), Olivier Teytaud[2], and Shi-Jim Yen[3]

[1] CSE, The University of New South Wales, Sydney, Australia
abdallahs@cse.unsw.edu.au
[2] Tao team, Inria, LRI, University of Paris-Sud, Orsay, France
[3] Ailab, CSIE, National Dong Hwa University, Shoufeng Township, Taiwan

Abstract. The game of Go is often said to be EXPTIME-complete. The result refers to classical Go under Japanese rules, but many variants of the problem exist and affect the complexity. We survey what is known on the computational complexity of Go and highlight challenging open problems. We also propose a few new results. In particular, we show that Atari-Go is PSPACE-complete and that hardness results for classical Go carry over to their Partially Observable variant.

1 Introduction

The 2000's and 2010's have seen a huge progress in computer Go, with the advent of Monte Carlo Tree Search (MCTS) [3], sequence-like simulations [7], rapid action value estimates [6], and human expertise included in the tree search. The game of Go is rich of terminologies; there are *semeais* which are hard for MCTS algorithms, *ladders* which connect remote parts of the board, *Ko fights* which are at the heart of Robson's proof of EXPTIME-hardness for Go with Japanese rules, and *Tsumegos* (problems of varying difficulty, studied both by beginners and experts). It is also rich of variants, such as Killall Go and the partially observable variant *Phantom Go*.

The complexity of problems for a game is usually formalized as follows. A position is any particular state of the game, i.e., in the case of Go with Japanese rules a board equipped with stones. We consider a family P of initial positions, and we consider the following decision problem.

Sure win problem: Is a given position $p \in P$ a sure win for Black assuming perfect play?

The complexity of a game is the complexity of this decision problem. It is well defined for fully observable (FO) games, as long as rules are formally described, and all information pertaining to a state is included in the input. We have

- For Go with Japanese rules, not all situations are completely formalized. Fortunately, the rules are not ambiguous for a large subset of positions, and we can still work on the complexity of Go with Japanese rules.
- Chinese rules present a distinct problem: some necessary information is not included in the position. Indeed, because cycles are forbidden by the

© Springer International Publishing Switzerland 2015
A. Plaat et al. (Eds.): ACG 2015, LNCS 9525, pp. 76–88, 2015.
DOI: 10.1007/978-3-319-27992-3_8

superko rule, one must remember past positions and avoid them. We will consider the decision problem associated to positions, considering there is no history of forbidden past state. This is the classical considered setting, e.g., in [8].

For partially observable variants, we still assume a fully observed input state, which is then played with its partially observable setting. The decision problem is not a sure win anymore but rather winning with a sufficiently high likelihood.

> **Threshold problem:** Given a fully observable position $p \in P$ and a rational number $c \in [0, 1]$, is there a strategy for Black (possibly but not necessarily randomized) so that for all White strategies, Black wins with probability at least c when the game starts in position p?

2 Rules, Variants, and Terminology

At the beginning of the game, the board is empty. Players, starting with Black, take turns putting a stone of their color on a free intersection. When a maximum group of 4-connected stones of a same color has no *liberty* (i.e., does not touch any free intersection), then it is removed from the board — this is termed *capture*. Suicide, i.e., playing a stone that would be immediately captured, is forbidden. Playing a move which goes back to the previous situation is forbidden; this is the *Ko* rule. In some versions, any move which goes back to an earlier position is forbidden; this is the *Superko* rule. The game terminates when no players want/can play. The *score* of a player, according to Chinese rules, is then the number of stones on the board, plus the surrounded empty locations. As Black plays first, White is given a bonus, termed *Komi*. The player with the greatest score is the winner.

Black's inherent first move advantage may skew games between equally skilled opponents. It is therefore traditional to give White a number of bonus points called *Komi* as a compensation. The larger the Komi, the easier the game for White. A Komi of 7 is considered fair, but to prevent draws, non-integer Komi such as 6.5 and 7.5 are regularly used. Games between unequally skilled opponents need to use a balancing scheme to provide a challenge for both players. The *handicap* mechanism is popular whereby Black is given a specified number of free moves at the start of the game, before White plays any move. The larger the handicap, the easier the game for Black.

Chinese and Japanese Rules. In most practical settings, there is no difference between using Chinese or Japanese rules. In mathematical proofs, however, it makes a big difference, as the latter allow cycles. Cycles of length 2 (termed *ko*) are forbidden in all variants of the rules, but longer cycles (termed *superko*) are forbidden in Chinese rules, whereas they are allowed in Japanese rules.

In Japanese rules, when a cycle occurs, the game is considered as a draw and replayed.[1] More than 20 pro games draw with superko are known in the

---

[1] See the detailed rules at http://www.cs.cmu.edu/~wjh/go/rules/Japanese.html.

world since 1998. On 2012, September 5th, in the Samsung world cup, the game Lee Shihshi - Ku Li featured a quadruple Ko, leading to a loop in the game, which was replayed.

Ladders are an important family of positions in Go. They involve a growing group of stones, e.g., white stones; Black is trying to kill that group, and White is trying to keep it alive by extending it. At each move, each player has a limited set of moves (one or two moves usually) which are not immediate losses of the ladder fight. An interesting result is that ladders are PSPACE-hard [9]: one can encode geography (PSPACE-complete) in planar-geography, and planar-geography in Go. So, Go, even restricted to ladders, is PSPACE-hard. This result is less impressive than the EXPTIME-hardness, but the proof is based on moderate size gadgets which can almost really be reproduced in a Go journal.

[4] proved that *Tsumegos*, in the restricted sense of positions in which, for each move until the problem is over, one player has only one meaningful move and the other has two, are NP-complete. This means that there is a family of Tsumegos (with the previous definition, from [4]) for which the set of positions which are a win for Black, in case of perfect play, is NP-complete.

2.1   Go Variants

Killall is a variant of Go in which Black tries to kill all groups of the opponent and White tries to survive. The Killall Go variant can be seen as a special case of Classic Go with appropriately set Komi and handicap. The Komi is almost the size of the board, so that White wins if and only if she[2] has at least one stone alive at the end of the game. The handicap is large enough that the game is approximately fair. For example, on a 9×9 board, the Komi would be set to 80.5 and the handicap would be 4 stones; on a 19×19 board, the Komi would be 360.5 and the handicap between 17 and 20 stones.

Atari Go is another variant, in which the first capture leads to a victory. Therefore, Komi has no impact. Atari Go is usually played as a first step for learning Go.

Phantom Go is a partially observable variant of Go. Given a fully observable (FO) game G, *Phantom G* denotes the following variant. A referee (possibly a computer is the referee) takes care of the board, and players only see their stones on their private version of the board. That is, White is only informed of the position of White stones, whereas Black is only informed of the position of Black stones. Moves are not announced, so even with perfect memorization, it is usually not possible to know where the opponent stones are. Illegal moves should be replayed. Depending on variants, the player can get additional information, such as captures, and possibly the location of captured stones.

Blind Go exactly follows the rules of Go, except that the board is not visible; players must keep the board in mind. In case of an illegal move due to a memorization error, they should propose another move. *One-color Go* is an easier version: there are stones on the board, but they all have the same color;

---

[2] For brevity, we use 'she' and 'her' whenever 'she or he' and 'her or his' are meant.

players should memorize the "true" color of stones. In these variants, the computational complexity is the same as in the original game; contrary to Phantom Go, the difference with Go boils down to a memory exercise.

3   Results in Fully Observable Variants

Below we discuss the results in fully observable variants of Go. We do so for the Japanese rules (3.1), the Chinese rules (3.2), the Killall Go variant, and Atari Go (3.4).

3.1   Japanese Rules

[10] has proved the EXPTIME-hardness of a family P of Go positions for which the rules are fully formalized. This is usually stated as EXPTIME-completeness of Go. The point is that Go with Japanese rules is not formally defined; there are cases in which the result of the game is based on precedents or on a referee decision rather than on rules (confusing examples are given in http://senseis.xmp. net/?RuleDisputesInvolvingGoSeigen, and precedents are discussed in http:// denisfeldmann.fr/rules.htm#p4).

However, FO games are EXPTIME-complete in general, unless there are tricky elements in the evaluation of the result of the game (more formally: if evaluating a final state cannot be done in exponential time) or in the result of a move (more formally: the board after a move is played can be computed in exponential time); therefore, we can consider that Go is EXPTIME-complete for any "reasonable" instantiation of the Japanese rules, i.e., an instantiation in which (i) the situations for which the rules are clear are correctly handled and (ii) deciding the consequences of a move and who has won when the game halts can be done in exponential time.

The original proof by Robson is based on Ko fights, combined with a complex set of gadgets that correspond to a ladder [10]. More recent work has shown that these ladder gadgets could be simplified [5].

3.2   Chinese Rules

Go with Chinese rules is different. The same Ko fight as in Robson's proof can be encoded in Chinese rules, but the result is different from that by the Japanese rules (at least if humans follow the superko rules forbidding cycles, which is not that easy in some cases). In contrast, the PSPACE-hardness is applicable with Chinese rules as well as with Japanese rules; therefore, one might, at first view, believe that Go with Chinese rules is either PSPACE-complete or EXPTIME-complete. However, the state space with Chinese rules is much bigger than the size of the apparent board: one must keep in memory all past positions, to allow avoiding cycles. As a consequence, Go with Chinese rules is not subject to the general EXPTIME result; and EXPSPACE is the current best upper bound.

Theorem 1 (Folklore Result). *Go with Chinese rules is in* EXPSPACE.

Proof. Extend the state with an (exponential size) archive of visited states. This augmented game is acyclic. Therefore it is solved in polynomial space (by depth first search of the minimax tree) on this exponential size representation; this is therefore an EXPSPACE problem.

This is widely known and we do not claim this as a new result. Lower bounds for games with a no-repeat condition are typically much more difficult to obtain, but the general case is EXPSPACE-hard [8, Sect. 6.3]. There is no EXPTIME-hardness of EXPSPACE-hardness result for the specific case of Go with Chinese rules. So the actual complexity of Go with Chinese rules is open and might lie anywhere between PSPACE and EXPSPACE. A nice consequence in complexity theory is that if Go with Japanese rules is harder than Go with Chinese rules (in a computational complexity perspective), then EXPTIME (where we find Go with Japanese rules) is different from PSPACE.

3.3   Killall Go Variant

A detailed proof of complexity for Killall Go is beyond the scope of this paper; but we give a few hints in this direction. A key component for applying classical results (such as the EXPTIME-hardness with Japanese rules or the PSPACE-hardness with Chinese rules) is to rewrite

1. a big group, the life or death of which decides the result of the game (this is a key component for all proofs); this big group can make life only through a ladder;
2. the Ko-fight (necessary for the EXPTIME-hardness of Japanese Go);
3. the ladder components from [5], or the ones from [10] (both are equivalent, the ones from [5] are simpler), necessary for both the PSPACE-hardness of Chinese Go and the EXPTIME-hardness of Japanese Go.

These components should be adapted to the Killall Go setting. The first part of this, number 1 above, consists in designing a position in which winning the game boils down to winning the ladder, thanks to a big group which must live thanks to the ladder. First, we need a group which will live or die, only depending on the ladder. This is easily obtained as shown in Fig. 1.

The Ko fight is also not a problem; there is no room for making life around the Ko fight, so Killall Go and Go are equivalent for this part of the game. Then, we must adapt the gadgets for the ladder itself. The difficult point is to ensure that there is no room elsewhere on the board in which White might make life, out of the line of play used in the widgets. The principles are as follows.

– Fill all empty spaces with strong Black stones, which cannot be killed. We only have to keep two empty points beside each point of the ladder path; we can fill all other parts of the board with black stones and their eyes.
– Since the ladder path is thin and the surrounding Black stones are very strong, White cannot make two eyes even if she plays three continuous white stones in a row. We get ladders as in, e.g., Fig. 2.

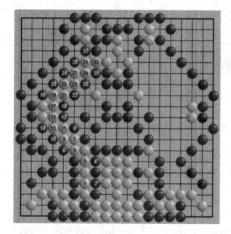 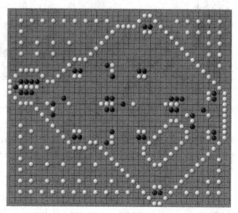

Fig. 1. A position in which White wins (in a Killall Go sense) if and only if she can win the ladder.

Fig. 2. White to play. A ladder, adapted to Killall Go: the left-most Black group can make life by killing in this ladder. Black cannot make life without winning the ladder.

We believe this statement can be made rigorous by a long and tedious formalization; this is however a very long exercise and beyond the scope of this work; so, we only have the following conjecture.

Conjecture 1. Killall Go with Japanese rules is EXPTIME-hard. Killall Go with Chinese rules is PSPACE-hard and EXPSPACE.

Note that Killall Go with Chinese (resp. Japanese) rules is in EXPSPACE (resp. EXPTIME) because it forms a subset of decision problems for Classical Go.

3.4   Atari Go

There is no capture until the end of a game of Atari-Go and each turn a new stone is added to the board, therefore the game is polynomially bounded and is in PSPACE. To prove hardness, we will reduce from the two-player game Generalized Geography. The variant we use assumes a planar bipartite graph of degree at most 3 and is PSPACE-hard [9].

We can partition the set of vertices of a planar bipartite graph of degree 3 based on the indegree, the outdegree, and the color of the vertex. We can restrict ourselves to considering the following 6 types of vertices: White vertex with indegree 1 and outdegree 1 (W1-1 vertex for short), W2-1, W1-2, B1-1, B2-1, and B1-2. The Atari Go gadgets for each White vertex type are presented in Fig. 3, and the gadgets for Black vertex types are the same with the color reversed.

The gadgets comprise *interior groups*, *exterior groups*, and *link groups*. The interior groups only appear in 1-2-vertex gadgets and initially contain 1 or

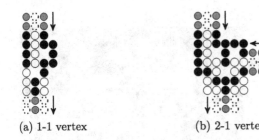
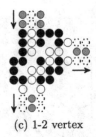

(a) 1-1 vertex   (b) 2-1 vertex   (c) 1-2 vertex

Fig. 3. Gadgets reducing Generalized Geography on bipartite planar graphs of degree at most 3 to Atari-Go. Each gadget represents a type of white vertex, the black vertices can be represented by flippling colors.

3 stones. The exterior groups all have a large number of liberties and will never be under any capturing threat. The link groups serve as links between the different gadgets. Each link group is flanked on each side by an exterior group of the opposite color, and the three groups take the shape of a corridor so as to imitate the path of the Generalized Geography edge. To each edge in the Generalized Geography instance corresponds a single link group in the Atari Go instance.

Our construction ensures that initially, each link group has two *end liberties* in the vertex gadget it arrives in and one *starting liberty* in the vertex gadget it departs from. As such, before the start of the game, every group on the board has at least 3 liberties, except for single-stone interior groups. Playing in the starting liberty of a link group simulates selecting the corresponding edge in Generalized Geography, we call it *attacking* this link group.

Lemma 1. *If the opponent just attacked a given link group, it is a dominant strategy for the player to extend this group by playing in one of the two end liberties.*

In the case of the 1-1 and 2-1 gadgets, one of the end liberties is only shared with an opponent's exterior group, so it is dominant to play in the other end liberty. This remark and Lemma 1 ensure that it is dominant for both players to continue simulating Generalized Geography on the Atari Go board.

Theorem 2. *Atari Go is* PSPACE-*complete.*

Proof. The branching factor and the game length are bounded by the size of the board so the problem belongs to PSPACE. Generalized Geography on planar bipartite graphs of degree 3 is PSPACE-hard and can be reduced to Atari Go as we have shown above. Therefore, Atari Go is PSPACE-hard as well.

While the complexity of Atari Go is the same as that of Ladders, a different proof was needed because one of the ladder gadgets involves an intermediate capture.

4   Phantom Go

Go is usually FO, but variants in which the opponent's moves are not observed have been defined. In order to preserve the consistency of games, when a move is illegal (typically, but not only, because it is played on an opponent's stone) then the player is informed, and the move is replayed. Detailed rules differ, but players are always informed when their stones are captured. A consequence is that Ko fights can be played as in the FO game.

Phantom Go is a nice challenge, in particular because the best algorithms are somehow simple; basically, the pseudo-code of the best algorithms for playing in a situation S is as follows, for some parameter N.

- For each possible move m, repeat N times:
    - let K be the number of unknown opponent stones;
    - randomly place the K unknown opponent stones on the board;
    - perform a Monte Carlo simulation.
- Play the move m with best average performance.

In particular, this is not a consistent approach; even with N infinite, such an algorithm is not optimal. Nonetheless it usually performs better than more sophisticated approaches [2].

4.1   Lower Bounds on Phantom Go Complexity

We here present lower bounds on Phantom Go complexities derived by adapting the proofs in the FO case. There are two sets of gadgets we would like to use in the phantom framework.

- The Ko fight defined in Robson's work [10]. The situation here is easy, because, in Phantom Go rules, captures are visible. Ko fights involve one capture per move; therefore, they are played exactly as if it was standard Go.
- The ladder gadgets, either from [10] or [5]. These gadgets are necessary both for adapting Robson's proof to the Phantom Go case (there is more than a Ko fight in [10], there is also a ladder), and directly for the PSPACE-hardness of ladders in the Phantom Go case. We reproduce the gadget ladders in Fig. 4 for convenience [5, Fig. 2].

Theorem 3. *Ladders are* PSPACE-*hard also in Phantom Go with both Chinese and Japanese rules. Go is* EXPTIME-*hard in Phantom Go with Japanese rules.*

This holds for the existence of a sure win for Black, and therefore also for the threshold problem.

Proof. We consider the problem of the existence of a sure win for Black.

Preliminary Remark: Black has a sure win, if and only if Black has a deterministic strategy for winning surely against the White player playing uniformly at random. Therefore, Black has a sure win even if the White player, by chance, plays exactly as if White could see the Board.

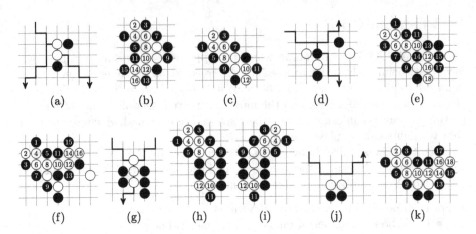

Fig. 4. Gadgets to build a PSPACE-complete family of ladder problems [5, Fig. 2]. We show that the same family of ladder problems is PSPACE-complete in Phantom Go.

The proof consists in using the same positions as in [5], and showing that strategies used in [5] can be adapted to the Phantom Go case, with the key property that optimal strategies in [5] are adapted to optimal policies in the Phantom Go case. This implies that we can adapt ladders in Phantom Go. Thanks to the remark above (Ko fights are played in Phantom Go with the same line of play as in Go) we can also play Ko fights. So, simulating ladders leads both to PSPACE-hardness (for both Chinese and Japanese rules) and to EXPTIME-hardness (for Japanese rules, because we simulate a Ko fight for which the hardness proof exists only in the case of Chinese rules).

So, let us consider ladders, and let us simulate them in Phantom Go. In the (fully observable) Go version of the game, the line of play is as follows.

Each player is forced to play the line of play described in Fig. 4, otherwise she loses the game [5]. The *Black Choice* Fig. 4a gadget leads to Fig. 4b or c, up to Black. *White Choice* Fig. 4d leads to Fig. 4e or f, up to White. *Join* leads to Fig. 4g, in case the ladder comes as in Fig. 4h or as in Fig. 4i. *Mirror* leads to Fig. 4k. In all cases, there are at most two choices for each player at each move.

Case 1: When White has a Forced Move. Let us consider the case in which, in Go, White has only one move M which is not an immediate loss. In the positions proposed in [5] this happens when the ladder just propagates, without any choice, including the mirror and the join.

Let us show that it is also the case in Phantom Go. This case is the easiest: the killing move for Black is precisely the move M that White should play. So Black can just try to play M. If M kills, Black wins. Otherwise, Black is informed that White has played M, since the rules of Phantom Go state that in case a move is illegal for Go it should be replayed elsewhere.

Case 2: When Black has a Forced Move, i.e., a Move Such that All Other Moves Immediately Make the White Ladder Alive. By the preliminary remark, if Black does not play this move, it does not have a sure Win. This case is concluded.

Case 3: When White has a Choice. Let us consider the case in which, in Go, White has two choices. This is the case in "White Choice" gadgets. Let us show that it is also the case in Phantom Go, and that Black can guess which move White has played. Let us consider that White plays one of the two choices displayed in Fig. 4e and f (Fig. 4, move 12 in Fig. 4e and f respectively). Let us show that Black can guess which move and reply as in Fig. 4e and f, even in the Phantom Go case. This is easy because the move number 12 in Fig. 4e is a capture. So if White plays this move, Black is informed, and we are back at the line of play of Fig. 4e. Otherwise, Black can just check that White has played the move 12 in Fig. 4f by trying to play it: if White had not actually played that move, then this captures the ladder and Black has won.

Case 4: When Black has a Choice. Let us consider the case in which, in the FO line of play, Black has two choices. By the preliminary remark, Black wins surely only if it wins surely if White plays with knowledge of Black's move, so we can consider the case of omniscient White. If Black does not play one of the two moves, White makes life in the FO line of play. If Black plays one of the two moves, we can consider the case in which White has made a correct assumption on which move Black has played.

4.2   Upper Bounds on Phantom Go Complexity

Japanese Rules. Given that some partially observable deterministic two-player games can be undecidable even when restricted to a finite state space [1], it is not even clear that Phantom Go with Japanese rules is decidable at all. The best bound we have is the following.

Theorem 4. *Phantom Go with Japanese rules is in 0', the set of decision problems which can be decided by a Turing machine using the halting problem as an oracle.*

Proof. Thanks to Theorem 5 in [1], the following machine halts if and only if Black has a strategy for winning with probability greater than c, in Phantom Go with Japanese rules.

- $K = 1$
- while (true)
  - Solve exactly the game up to K time steps, starting in a given position p.
  - if Black can win with probability $> c$, then halt.
  - Otherwise, $K \leftarrow 10K$

Then, the following machine with the halting problem as oracle can solve the decision problem of the present theorem.

- Input: a position p.
- Output:
  - output *yes* if the machine above halts.
  - output *no* otherwise.

This yields the expected result.

It implies that finding the optimal move is also computable in 0'. So, as a summary: Phantom Go with Japanese rules is EXPTIME-hard and in 0'.

Chinese Rules. We propose an upper bound for Phantom Go, in the case of Chinese rules. Due to Chinese rules, there is a superko rule: it is forbidden to have twice the same move. We claim the following.

Theorem 5. *Phantom Go with Chinese rules is in* 3EXP, *the set of decision problems which can be solved in* $2^{2^{2^{\text{poly}(n)}}}$ *time, where* $\text{poly}(n)$ *is a polynomial function.*

Proof. We consider games of Phantom Go with Chinese rules, on an $n \times n$ board, and we compute upper bounds on the number of board positions, B, the maximum length of a game, L, the number of distinct histories, N, and the number of pure (i.e., deterministic) strategies, P. We first have $B = 3^{n \times n}$, since any intersection can be empty, black, or white.

The length of a game is at most $L = 3(1 + n \times n)B$, where the 3 factor indicates whether players have passed zero, one, or two times in the current position. The $n \times n + 1$ factor arises because a player can play again when their move is rejected in Phantom Go, for each "real" move, there are at most $n \times n + 1$ trials (including pass), which can be (i) accepted or (ii) rejected as illegal. Finally the B factor is due to the superko rule: each position is allowed at most once.

One move leads to $(n \times n + 1)$ possibilities (including pass). Then, the player observes some information, which is of size at most $1 + 2^{n \times n}$; this is the number of possible captures, plus one for the "illegal" information. Therefore, an upper bound on the number of histories is $N = ((n \times n + 1)(1 + 2^{n \times n}))^L$.

Pure strategies are mappings from histories to actions; there are therefore at most $P = (1 + n \times n)^N$ pure strategies. The $1 + n \times n$ stands for the $n \times n$ standard moves, plus a possibility of pass.

Substituting the B, L, and N for their values, we obtain that the number of pure strategies is at most a tower of 3 exponentials in n. Solving the input Phantom Go position can therefore be reduced to solving a normal-form game of size triply exponential in n. Two-player zero-sum games in normal-form can be solved in time polynomial in the size of the matrix, for instance by solving the corresponding linear program, so Phantom Go can be solved in time triply exponential in n.

Table 1. Summary of the complexity of Go-related problems. Results with a * have not been fully formalized yet and should be considered as conjecture only. New results are highlighted in boldface.

Rules	Variant	Lower bound	Upper bound
	Atari-Go	PSPACE-**hard**	PSPACE
Chinese	Classic	PSPACE-hard	EXPSPACE
	Phantom	PSPACE-**hard**	**3exp**
	Killall	PSPACE-hard*	EXPSPACE
Japanese	Classic	EXPTIME-hard	EXPTIME
	Phantom	EXPTIME-**hard**	**0'** (decidability unsure)
	Killall	EXPTIME-hard*	EXPTIME

5 Conclusion

We (i) surveyed the state of the art in the complexity of Go, (ii) proved a new result for Atari-Go, (iii) proposed (without formal proof) the extensions of these results to Killall Go variants, and (iv) proved lower and upper bounds for Phantom Go. The main open problems are (i) the decidability of Phantom Go with Japanese rules (because undecidability would be the first such result for a game played by humans), (ii) the complexity of Go with Chinese rules (because the gap between the lower and upper bounds is huge, more than for any other game), and (iii) the complexity of Phantom Go with Chinese rules. Table 1 summarizes the known results on the complexity of Go and its variants. Upper bounds for Japanese rules assume that we only consider reasonable formalizations of the rules and rule out ambiguous positions.

Acknowledgments. We are grateful to Tristan Cazenave for fruitful discussions around Phantom Go. The first author was supported by the Australian Research Council (project DE 150101351).

References

1. Auger, D., Teytaud, O.: The frontier of decidability in partially observable recursive games. Int. J. Found. Comput. Sci. **23**(7), 1439–1450 (2012)
2. Cazenave, T., Borsboom, J.: Golois wins Phantom Go tournament. ICGA J. **30**(3), 165–166 (2007)
3. Coulom, R.: Efficient selectivity and backup operators in monte-carlo tree search. In: van den Herik, H.J., Ciancarini, P., Donkers, H.H.L.M.J. (eds.) CG 2006. LNCS, vol. 4630, pp. 72–83. Springer, Heidelberg (2007)
4. Crâsmaru, M.: On the complexity of Tsume-Go. In: van den Herik, H.J., Iida, H. (eds.) CG 1998. LNCS, vol. 1558, p. 222. Springer, Heidelberg (1999)
5. Crâsmaru, M., Tromp, J.: Ladders are PSPACE-complete. In: Marsland, T., Frank, I. (eds.) CG 2001. LNCS, vol. 2063, p. 241. Springer, Heidelberg (2002)

6. Gelly, S., Silver, D.: Combining online and offline knowledge in UCT. In: Proceedings of the 24th International Conference on Machine Learning, ICML '07, pp. 273–280. ACM Press, New York (2007)

7. Gelly, S., Wang, Y., Munos, R., Teytaud, O.: Modification of UCT with patterns in Monte-Carlo Go. Rapport de recherche INRIA RR-6062 (2006). http://hal.inria.fr/inria-00117266/en/

8. Hearn, R.A., Demaine, E.D.: Games, Puzzles, and Computation. A K Peters, Cambridge (2009)

9. Lichtenstein, D., Sipser, M.: Go is polynomial-space hard. J. ACM **27**(2), 393–401 (1980)

10. Robson, J.M.: The complexity of Go. In: IFIP Congress, pp. 413–417 (1983)

On Some Evacuation Games
with Random Walks

Matthias Beckmann[(✉)]

University of Jena, Jena, Germany
matthias.beckmann@uni-jena.de

Abstract. We consider a single-player game where a particle on a board has to be steered to evacuation cells. The actor has no direct control over this particle but may indirectly influence the movement of the particle by blockades. We examine optimal blocking strategies and the recurrence property experimentally and conclude that the random walk of our game is recurrent. Furthermore, we are interested in the average time in which an evacuation cell is reached.

1 Introduction

Our point of interest is a special class of random walk games. Whilst in simple discrete random walks every neighboring cell is reached with probability one divided by the number of neighboring cells, our model discusses a random walk with a slight control component. In every single step of the random walk, a part of its connections to neighboring cells may be blocked. The intention is to guide the particle of this random walk to an evacuation cell. The optimal blocking strategy can be derived by solving a non-linear equation system. A related model is discussed in [2].

We can also find a similar blocking principle in the thought experiment of the Maxwell Demon [5]. The Maxwell Demon sorts particles into slow moving and fast moving kinds by opening or blocking a passage between two rooms, depending on what kind of particle is located in front of the passage.

The motivation for this game came up when a bumblebee got lost in Ingo Althöfer's living room and the question was how to get the insect out of the room without inflicting injuries to her[1] [1]. The seemingly erratic moving pattern of the bumblebee reminds one of a random walk.

Polya's Theorem [6] is an important result for simple random walks. It is often paraphrased as "A drunk man always finds his way home, but the drunk bird may not find its way home". The concrete mathematical formulation is that the simple random walk returns to the point of origin with probability one in dimensions one and two, while in higher dimensions this probability is lower than one. We will examine this property experimentally for our random walk game and observe that the random walk in our model is recurrent in all dimensions.

---

[1] For brevity, we use 'he' and 'his', whenever 'he or she' and 'his or her' are meant.

© Springer International Publishing Switzerland 2015
A. Plaat et al. (Eds.): ACG 2015, LNCS 9525, pp. 89–99, 2015.
DOI: 10.1007/978-3-319-27992-3_9

This paper is thematically divided into two parts. The first part concerns the basic game and computation of optimal strategies. It consists of Sects. 2 and 3. The Basic Game and its applications are outlined in Sect. 2. We demonstrate how to compute optimal blocking strategies in Sect. 3 and show results for three medium size boards graphically. In the second part, consisting of Sects. 4, 5 and 6, we cover the time aspect of optimal strategies derived in the first part as well as a new simple heuristic.

2   The Basic Game

For the sake of easier reading we limit our description to the two dimensional game. We assume that the reader may derive the corresponding definitions for higher dimensional games easily and apply them to the higher dimensional games in later parts of this paper.

We consider a single particle on a board. Important examples are the infinite board $\mathbb{Z}^2$ and finite boards $\{1, 2, \ldots, m\} \times \{1, 2, \ldots, n\}$. The goal is that our particle reaches an evacuation cell as soon as possible. Evacuation cells are distinguished cells of the board. If the particle reaches such a cell, the game ends. Time in this game passes in discrete steps. In every step the particle moves into one of the permitted neighbor cells. The user may block the access to one specific neighboring cell. In further extensions of the model the access to more than one neighboring cell is blocked. Usually the number of allowed blockade placements is much lower than the number of neighbors.

In addition to that we forbid all blocking strategies which would corner the particle on its current cell. This is especially important for finite boards.

The process of blocking over one time step is shown graphically in Fig. 1. Figure 1(a) shows an initial situation for our model. The moving particle is represented by the circle denoted with a B. The objective is to reach the grayed out evacuation cell on top of the board. We can think of B as a bumblebee and the 5×5 board representing Ingo's living room. Outdoors is modeled by the evacuation cell and the window of our room, leading to outdoors, is represented by a small white rectangle.

Figure 1(b) shows the placement of a single blockade. In this case the blockade is placed south of the bumblebee. The remaining three possible moves are shown in Fig. 1(c). Each of them has a probability of $\frac{1}{3}$. Figure 1(d) displays one of the possible follow-up positions. This position would be the initial position for the next time step.

We imagine multiple real world applications of our model. For one we may interpret the board as a top view of a room, the particle as a bumblebee which got lost in Ingo's room and the window through which the bumblebee shall escape is represented through evacuation cells. A human now tries to encourage the bumblebee to move to the window by blocking its flight path with a piece of paper.

In a second approach we can view the board as a bird's eye view of a city with a grid-like street network. The cells in our model represent intersections of these streets. In our city there is an unruly mob, for example, upset soccer fans

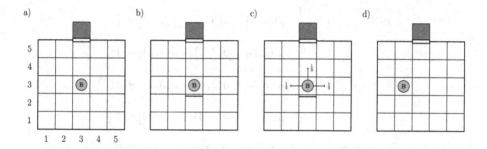

Fig. 1. A single time step in the basic model. Here for particle B the southern direction is blocked.

moving in a random pattern through the streets. The goal is to evacuate this mob to a certain point for example a train station. The means of blocking in this model is a police team which may block one street per time unit. The goal is to get the mob to the evacuation point as soon as possible. The decider is the head of the police who is looking for the optimal use of her police officers.

Thirdly we can apply the game to a special atomic diffusion problem. We consider a crystal with a single particle inside. The task is to move the particle out of the crystal and thus make the crystal flawless. We want to use diffusion to move the particle out of the crystal. To start the random walk process the crystal is heated up. Blockades may be realized through uneven heating or with the help of lasers.

3 Computing Optimal Strategies and Results

For every non-evacuation cell (i, j) on our board we define

$$a(i, j) = \text{Expected number of steps till evacuation}$$
$$\text{if an optimal blocking strategy is used.}$$

In addition we set $a(i, j) = 0$ for every evacuation cell (i, j).
For every non-evacuation cell (i, j) the following equations hold

$$a(i, j) = \min\{\text{Expected value of the blocking options} + 1\}$$

The equation system of the example from Fig. 1 consists of 26 equations, 25 for the cells in the room and one trivial equation for the evacuation cell. To make understanding the equation system easier, we give the equations for three selected cells $(3, 3)$, $(1, 1)$ and $(5, 3)$.

The cell $(3, 3)$ is located in the interior of the board. Here we are allowed to place the blockade in one of the positions *west, east, north* or *south*. Thus, for $a(3, 3)$ the equation

$$a(3,3) = \min \left\{ \frac{1}{3} \cdot \left[a(4,3) + a(3,4) + a(2,3) \right] + 1; \right.$$
$$\frac{1}{3} \cdot \left[a(2,3) + a(3,2) + a(4,3) \right] + 1;$$
$$\frac{1}{3} \cdot \left[a(3,4) + a(2,3) + a(3,2) \right] + 1;$$
$$\left. \frac{1}{3} \cdot \left[a(3,2) + a(4,3) + a(3,4) \right] + 1 \right\}$$

holds. Here the first term in the minimum operator represents the option of blocking *west*. $\frac{1}{3} \cdot \left[a(4,3) + a(3,4) + a(2,3) \right]$ is the average number of steps of the three possible successor cells after the step. The additional one counts the step from the actual cell to the new cell.

In the corner of the board, the movement options as well as the blocking options are limited by the boundaries of the board. Exemplarily for the lower left corner cell $(1,1)$ we are only allowed to put a blockade in the positions *north* or *east*. Thus the equation

$$a(1,1) = \min \left\{ a(1,2) + 1; a(2,1) + 1 \right\}$$

holds.

Similarly, for the cell $(5,3)$ on the boundary, the equation

$$a(5,3) = \min \left\{ \frac{1}{2} \cdot \left[a(5,4) + a(5,2) \right] + 1; \right.$$
$$\frac{1}{2} \cdot \left[a(4,3) + a(5,2) \right] + 1;$$
$$\left. \frac{1}{2} \cdot \left[a(5,4) + a(4,3) \right] + 1 \right\}$$

holds.

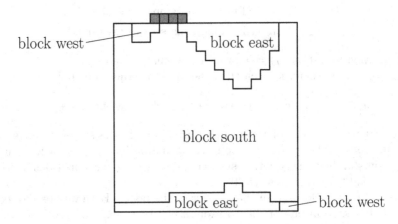

Fig. 2. Optimal strategy for a special 20×20 board. The board gets partitioned into five areas. In all cells of an area the blocking direction is identical.

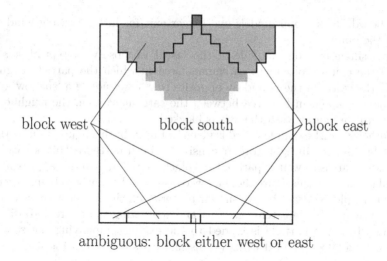

Fig. 3. Optimal strategy for a special 21×21 board. There is a single cell in the southern part where the optimal strategy is ambiguous.

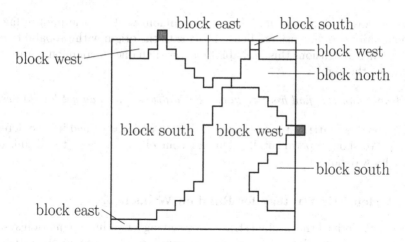

Fig. 4. Optimal strategy for a 20×20 board with two exits. The board can roughly be divided into two parts. In each of these parts the controller steers the particle towards the exit in this part.

We can compute the solution for this system of non-linear equations numerically. We obtain the optimal strategy used by memorizing where the minimum was reached. While our equation system has a unique solution the optimal strategy does not have to be unique. (See the Appendix.) Typical examples with multiple optimal strategies are symmetric boards.

Figure 2 shows the optimal strategy for a 20×20 board with four additional evacuation cells. Possible blockade options for each cell are *west, east, north* and *south*. The four evacuation cells are represented by gray cells on the upper ledge of our board. This group models one bigger exit for instance a big window in case of the bumblebee.

For a single exit on one side of the board we observe a typical strategy. Around the exit we observe a catchment area in which the particle is guided towards the exit. On the boundary opposite to exit we notice a 'shadow' of the catchment area and in the area between the catchment area, the neighbor cell opposite to the border with the exit is blocked.

Another 21×21 example board is given in Fig. 3. In this figure we underlayed the catchment area in light gray. It consists of both northern "block east" and "block west" areas as well as part of the "block south" area and is approximately of circular shape. In this figure the shadow consists of the southern boundary cells.

For multiple exits we observe similar patterns which interfer with each other, limiting the size of the catchment area and seperating the board into different zones in which the particle is guided to the exit accompanying the zone. An example for a 20×20 board with two distinct exits is given in Fig. 4.

4   Evacuation Models and Recurrence

Polya's theorem [6] states that the simple random walk on the infinite line or on the infinite two dimensional board returns to the origin with probability one, but in higher dimensions this probability is less than one. A common paraphrase for Polya's theorem is

"A drunk man will find his way home, but a drunk bird may get lost forever."

This phrase is attributed to Shizuo Kakutani and was coined in a conference talk [3]. We show experimentally that our control aspect helps the drunk bird to find its way home.

4.1   A Heuristic Strategy for Random Walks in $\mathbb{Z}^n$

In Sect. 3 we derived the optimal strategy by solving a non-linear equation system numerically. It is possible to define a corresponding infinite equation system for the infinite board $\mathbb{Z}^n$ and solve it numerically. But in this case it is not possible to solve it via our method described in the appendix.

We propose a simple *heuristic strategy* for our infinite board with one blockade and a single evacuation cell in the following way.

Strategy: For each cell block the access to the neighboring cell with the largest component.

Two-dimensional example: Our goal is at $(0, 0)$ and the particle is in the cell $(2, 4)$ with neighboring cells $(1, 4), (3, 4), (1, 3), (2, 5)$. Here the access to $(2, 5)$ would be blocked.

In some cases there is more than one cell with the largest distance from the goal in one component. We break the tie arbitrarily.

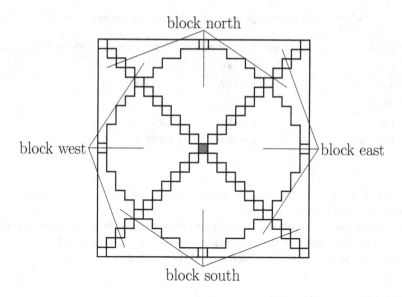

Fig. 5. Optimal strategy for a 23 × 23 board with a single evacuation cell in the center. Non-captioned cells have at least two optimal blockade positions.

This strategy is based on observations for the finite board. We calculated optimal strategies for square boards of side length 7, 11, 15, 19 and 23 and observed that the optimal strategy converges with growing board size to the simple strategy stated above. Figure 5 shows the optimal strategy for the 23 × 23 board with a single evacuation cell in the center. The circular area around the evacuation cell has the same strategy as the heuristic strategy. We observe that this area grows with increasing board size and always has a diameter of side length minus two. Thus, we expect that the optimal strategy on the infinite board is similar to the proposed strategy. While the heuristic strategy is not necessarily the optimal strategy, it is a good heuristic.

4.2 Simulation Results for the Heuristic Strategy on $\mathbb{Z}^n$

We estimate the expected number of steps until the random walk returns to the origin and the rate of return by simulating one million random walks for dimensions 2 to 4 and 100,000 random walks for dimensions 5 and 6. For each random walk 10,000,000 steps are simulated. If the random walk returns within this time frame we count this as a successful return. If it does not return to the origin we do not count it as a return. The results of the simulation are given in Table 1. The expected time till return to the origin is estimated by the arithmetic mean of the number of steps taken in our simulations. We also included the already known results for the simple random walk for comparision.

Remarkable is that the rate of return is one in all dimensions. Polya's theorem states that without blockades these rates are less than one for dimensions greater

Table 1. Recurrence property for the evacuation model on $\mathbb{Z}^n$

n	Expected time till return	Rate of return	Rate of return without blockades
2	11.5	1.00000	1
3	95.5	1.00000	0.340537
4	1020.7	1.00000	0.193206
5	13257	1.00000	0.135178
6	197890	1.00000	0.104715

than two. The rates for the simple random walk are given in the third column and were derived numerically by [4].

We also observe a connection between the dimension and the expected number of steps till return – if the dimension increases from n to $n+1$ the expected number of steps increases by a factor of about $2n+4$.

5 Evacuation Speed

In case of an evacuation it is important to know how fast our particle can reach an evacuation cell. For a certain cell on the board the number is given by the a variables from Sect. 2.

We are especially interested in boards of the following shape: The board is square with odd side length m. The only evacuation cell is located in the middle of one of the sides. One such board with $n = 2$ is shown in Fig. 1. A second board with $n = 10$ and the accompanying optimal blocking strategy is shown in Fig. 3. We want to examine how long it takes in average to evacuate the particle if it is placed randomly on one of the inner cells with probability $\frac{1}{m^2}$ for each cell. We examine this question for different numbers of blockades.

We choose just this class of boards in order to obtain a class of boards of growing size which only depends on a single parameter. The placement of the evacuation cell reflects our initial idea of evacuating a bumblebee out of a room.

To answer the question we define the *average time until evacuation* as

$$\overline{a_M} = \frac{1}{m^2} \sum_{\substack{(i,j) \text{ is an inner field} \\ \text{of the board}}} a(i,j),$$

where the set M indicates how many blockades may be placed at each cell, e.g., if $M = \{1, 2\}$ one or two blockades may be placed. The amount of allowed blockades might be different at corner and boundary cells because of the special rules for these cells outlined in the basic model. Thus on corner cells at most one access to a neighbor may be blocked and on boundary cells only two cells may be blocked at most.

We focus on the sets $M_1 = \{1\}$, $M_2 = \{1, 2\}$, and $M_3 = \{1, 2, 3\}$. In the case of M_3 all randomness is lost and the controller can always force the particle

Table 2. Expected time until evacuation for different blockade options

Blockade options	Expected number of steps $\overline{a_M}$
1	$2.2500 \cdot m + 20.3644$
1 to 2	$1.0000 \cdot m + 2.0015$
1 to 3	$\frac{3}{4} \cdot m + \frac{1}{2} + \frac{1}{4} \cdot \frac{1}{m}$

on the neighbor cell of her choice. Therefore, for a certain cell the time until evacuation is given by the Manhattan distance of this cell to the evacuation cell.

For sufficiently large boards with side length $m > 50$ we observe a linear relation between board size and average time until evacuation. The results for the optimal strategies of these blockade possibilities are shown in Table 2.

The results for M_1 and M_2 are computed by solving the corresponding equation systems. The result for M_3 was obtained by a short elementary computation.

6   Random Walk Half Life

In connection with Recurrence and Evacuation speed we are also interested in how long it takes to evacuate a particle with probability 50 %. In case of multiple repeated independent evacuations this property will reflect how much time is needed to save 50 % of a given population.

For this task we consider infinite boards of dimension n. Our particle starts at $(0,0)$ and the task is to let it return to this cell by placing a single blockade in every turn.

We estimate the Random Walk half life by Monte Carlo simulations. The numbers are given by the median of the time until return to the origin. The resulting half life periods for dimensions two to six are shown by Table 3.

Table 3. Experimental Results for the Random Walk half life time

n	Number of steps until evacuation of at least 50 %	Growth compared with $n-1$
2	6	–
3	34	5.67
4	532	15.65
5	8110	15.24
6	123691	15.25

We observe for dimension 3 or higher that if we increase the dimension by one, the number of steps until evacuation of at least 50 % increases by a factor of about 15.25.

7   Conclusions

We worked on two questions regarding the evacuation game. For question one we were concerned about how we should block the access to neighboring cells to get the object of concern in the fastest possible way to a predetermined cell. For question two we were interested in the time aspect of our evacuation.

For finite boards we can answer the question one of how to place blockades by solving a special equation system with min expressions. For the infinite board we derived a good heuristic strategy from our observations from the finite board. If we examine finite boards with a single exit we can divide the board into three areas where three blocking patterns are prevalent – a catchment area around the exit, a shadow opposite the exit and an intermediate area.

An important part of our investigation was whether the recurrence property holds or not in infinite boards of dimension 3 or higher. For the proposed simple heuristic we observed experimentally that the particle returns to the origin in all dimensions with probability one. In the context of the drunk bird parable we may conclude that just one blockade will help a drunk bird to find its home in any finite dimensions.

For question two, the expected time until a particle is evacuated, we observe a linear relation between side length of the 2-dimensional board under investigation and the evacuation time.

We also took a look at the time it takes to evacuate 50 % of a given population by using single blockades in each step. We observe that this time increases quickly with the dimension of the problem.

Research questions for the future are

- To find a formal proof for recurrence in high dimensions.
- Identifying optimal strategies on infinite boards in d dimensions.

Acknowledgments. The author would like to thank Ingo Althöfer for asking the interesting bumblebee evacuation question. Thanks also to three anonymous referees for their constructive comments.

Appendix

Method of Monotonous Iterations

Given an equation system (E) of the form $x = f(x)$ and a starting vector $x^{(0)}$ we compute $x^{(i+1)}$ by

$$x^{(i+1)} = f(x^{(i)}).$$

This method converges towards a solution, for instance, if the following conditions hold

1. (E) has a unique solution, and
2. the sequence $x^{(i)}$ is monotonically increasing in all coordinates and has an upper bound in each coordinate.

The first condition holds for the problems outlined in this paper.

The second condition depends on the starting vector $x^{(0)}$. Only a good starting vector will lead to the solution of (E). A good solution for our problems is $x^{(0)} = (0, 0, \ldots, 0)$ for which the second property is fulfilled.

References

1. Althöfer, I.: Personal Communication in June 2014
2. Althöfer, I., Beckmann M., Salzer. F.: On some random walk games with diffusion control (2015)
3. Durrett, R.: Probability: Theory and Examples (2010). http://www.math.duke.edu/rtd/PTE/PTE4_1.pdf. Accessed on 9th March 2015
4. Finch, S.R.: Polya's Random Walk Constant. Section 5.9 in Mathematical Constants, pp. 322–331. Cambridge University Press, Cambridge (2003)
5. Maxwell, J.C.: Theory of Heat, 9th edn. Longmans, London (1888)
6. Polya, G.: Ueber eine Aufgabe betreffend die Irrfahrt im Strassennetz. Math. Ann. **84**, 149–160 (1921)

Crystallization of Domineering Snowflakes

Jos W.H.M. Uiterwijk[✉]

Department of Knowledge Engineering (DKE),
Maastricht University, Maastricht, The Netherlands
uiterwijk@maastrichtuniversity.nl

Abstract. In this paper we present a combinatorial game-theoretic analysis of special Domineering positions. In particular we investigate complex positions that are aggregates of simpler fragments, linked via bridging squares.

We aim to extend two theorems that exploit the characteristic of an aggregate of two fragments having as game-theoretic value the sum of the values of the fragments. We investigate these theorems to deal with the case of multiple-connected networks with arbitrary number of fragments, possibly also including cycles.

As an application, we introduce an interesting, special Domineering position with value *2. We dub this position the *Snowflake*. We then show how from this fragment larger chains of Snowflakes can be built with known values, including flat networks of Snowflakes (a kind of *crystallization*).

1   Introduction

Domineering is a two-player perfect-information game invented by Göran Andersson around 1973. It was popularized to the general public in an article by Martin Gardner [9]. It can be played on any subset of a square lattice, though mostly it is restricted to rectangular $m \times n$ boards, where m denotes the number of rows and n the number of columns. The version introduced by Andersson and Gardner was the 8×8 board.

Play consists of the two players alternately placing a 1×2 tile (domino) on the board, where the first player may place the tile only in a vertical alignment, the second player only horizontally. The player first being unable to move loses the game; his opponent (who made the last move) is declared the winner. Since the board is gradually filled, i.e., Domineering is a converging game, the game always ends, and ties are impossible. With these rules the game belongs to the category of *combinatorial games*, for which a whole theory (the Combinatorial Game Theory, or CGT in short) has been developed. In CGT the first player conventionally is called Left, the second player Right. Yet, in our case we also sometimes will use the more convenient indications of Vertical and Horizontal, for the first and second player, respectively.

Among combinatorial game theorists Domineering received quite some attention. However, the attention was limited to rather small or irregular boards

© Springer International Publishing Switzerland 2015
A. Plaat et al. (Eds.): ACG 2015, LNCS 9525, pp. 100–112, 2015.
DOI: 10.1007/978-3-319-27992-3_10

[1,4,5,7,11,16]. Larger (rectangular) boards were solved using α-β search [12], leading to solving all boards up to the standard 8×8 board [2], later extended to the 9×9 board [10,15], and finally extended to larger boards up to 10×10 [6].

Recently the combination of CGT and α-β solvers was investigated [3]. First, endgame databases with CGT values for all Domineering positions up to 15 squares were constructed. Then they were used during the search process. Experiments showed reductions up to 99 %. As a sidetrack we discovered many interesting Domineering positions, some of which form the basis for the present paper.

2 Combinatorial Game Theory and Domineering

In this section we start by a short introduction to the field of CGT applied to Domineering, in Subsect. 2.1. We next describe games with number and nimber values in Subsect. 2.2; they are of particular interest for the remainder of this paper. Finally, we explain the notion of *disjunctive sums of games* in Subsect. 2.3. For a more detailed introduction we refer to [1,5,7], where the theory is explained and many combinatorial games including Domineering are treated.

2.1 Introduction to the Combinatorial Game Theory

In CGT applied to Domineering it is common to indicate the two players by Left (Vertical) and Right (Horizontal). A game is then described by its Left (vertical) and Right (horizontal) options, where options are again games. A Left option describes a legal move Left can make by giving the game position after performing the move, whereas a Right option describes a legal move which Right can make. To better understand this recursive definition, it is convenient to start by analyzing the simplest of games. A game of Domineering consisting of no or merely one square (see Fig. 1 left) has no legal move for either player, i.e., no Left options and no Right options, and is called zero or 0. In short, since games are represented by their Left and Right options, this game is described by $G = \{|\} = 0$.

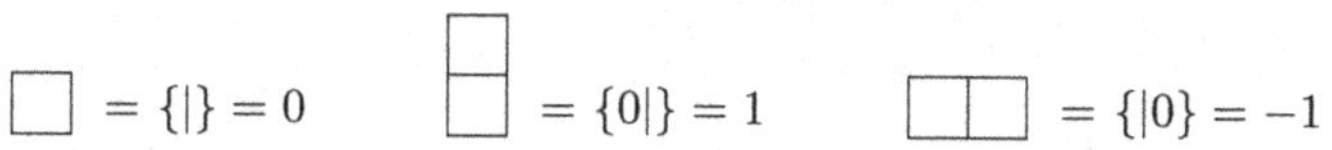

$$\square = \{|\} = 0 \qquad \begin{array}{c}\square\\\square\end{array} = \{0|\} = 1 \qquad \square\square = \{|0\} = -1$$

Fig. 1. From left to right, the simplest Domineering positions with values 0, 1, and -1.

The second and third game in Fig. 1 show the two next simplest Domineering games, 1 and -1; each have exactly one legal move for Left or Right, respectively. Using the previously shown game notation of Left and Right options, they are defined as $1 = \{\{|\}|\} = \{0|\}$ and $-1 = \{|\{|\}\} = \{|0\}$. By expanding the game board to larger and more arbitrary shapes, very complex representations can be obtained. Only a small portion of the larger games have easy values such as 0, 1 or -1. It is generally true that games with positive values pose an advantage for Left, and games with negative values pose an advantage for Right.

2.2 Number and Nimber Games

We next focus on two types of games with special characteristics, numbers and nimbers. As usual in CGT, we further freely use numbers and nimbers to indicate both a game and the value of that game. The context will make clear what is meant.

Numbers. The games 0, 1, and −1 have values that are called numbers. Numbers may be positive and negative integers, but can also be fractions. In games with finite game trees specifically, all numbers are *dyadic rationals*, i.e., rational numbers whose denominator is a power of two. Numbers are represented by the options for both Left and Right: $G = \{G^L | G^R\}$, where G^L and G^R are either $\{\} = \emptyset$ or sets of other numbers. For all numbers with two non-empty options it holds that $G^L < G^R$. Generally, the sign of a number indicates which player wins the game, while the value 0 indicates a win for the player who made the last move, i.e., the previous player. Numbers and combinations of multiple numbers thus have a predetermined outcome.

Infinitesimals. Infinitesimal games have values that are very close to zero and are less than every number with the same sign. Being infinitesimally close to zero, these games can never overpower any game with non-zero number value under optimal play. There are several types of infinitesimal values, such as ups/downs, tinies/minies, and nimbers. We next introduce the latter ones.

Nimbers. Nimbers commonly occur in impartial games, but may occur in partizan games like Domineering. Their main characteristic is that their Left and Right options are identical, following a certain schema, and that they are their own negatives. The only frequent representative of this group in Domineering, according to Drummond-Cole [8], is the star or ∗. In game notation a star looks as $G = \{0 | 0\} = *1 = *$. Figure 2 shows the most simple star position in Domineering.

Fig. 2. A Domineering star position and its Left and Right options.

The next number, ∗2, is defined as $*2 = \{0, * | 0, *\}$, and in general a nimber $*n$ is defined as $*n = \{0, *, ..., *(n-1) | 0, *, ..., *(n-1)\}$. Clearly, the first player to move in a nimber game wins by moving to 0, which is why the outcome is not predetermined and why its value is truly confused with zero. This introduces an interesting property of nimbers: every nimber is its own negative, i.e., $*n = -*n$. As a consequence, when two identical nimbers occur, they cancel each other: $*n + *n = 0$, where n is any integer ≥ 0.

When we add a third square to the central square of the position in Fig. 2, the value of the position does not change, since any move by a player necessarily destroys the central square, leaving unconnected squares with 0 as resulting sum value. This even holds when we add a fourth square to the central one. Therefore, the two positions Small-T and Plus in Fig. 3 also have value $*$.

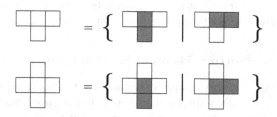

Fig. 3. Two more Domineering star positions and their Left and Right options, the Small-T (top) and the Plus (bottom).

2.3 Sums of Games

As a game progresses, it is common for many combinatorial games that the positions begin to decompose. This usually means that the game board consists of two or more components (subgames) that can no longer interact. Their independent nature allows for an independent analysis. Each subgame has a value and the combination of multiple components, denoted as the *sum of the subgames*, also has a value: the sum of their individual values.

More formally, the (*disjunctive*) sum of two games G and H is defined as $G + H = \{G^L + H, G + H^L \mid G^R + H, G + H^R\}$, with the superscripts indicating the Left or Right options of a game. The reasoning behind this definition is that a player can only move in one of the two subgames, thus to the Left or Right options of one game plus the entire other game.

When adding two numbers, the sum will always simplify back to a number again. In fact, number games and their values can be treated like mathematical numbers, and the sum is equivalent to their mathematical sum. When two nimbers are added, their sum is given by the nim-addition rule, and is a nimber again. Two equal nimbers add to $*0 = 0$, meaning that they cancel each other. Indeed, 0 is the only value being a number and a nimber at the same time. For any other type of game theoretical values the sum can easily become quite complex.

3   CGT Theorems for Domineering

In this section we focus on theorems in CGT, that are specifically applicable to Domineering positions. The first such theorem was already given by Conway in [7, p. 115]. We start this section by presenting Conway's theorem and proof, in Subsect. 3.1. This theorem is only applicable to linearly connected structures of exactly two fragments. In a previous publication [14] we extended this theorem to arbitrarily connected structures, involving up to four fragments. We repeat this theorem and its proof in Subsect. 3.2. In the present paper we extend this to structures of an arbitrary number of fragments (networks), possibly involving cycles. The new theorem and its proof are given in Subsect. 3.3.

3.1   The Bridge Splitting Theorem for Domineering

Though the preceding theory is applicable to any combinatorial game, Conway formulated a beautiful decomposition theorem that is specifically applicable to Domineering [7]. Because of its importance, we repeat the theorem and its proof. We denote it as the *Bridge Splitting Theorem*, to contrast it with the other theorems to be formulated in the next subsections.

Theorem 1 (Bridge Splitting Theorem). *If for some game $G\square$ its value is equal to that of the game G alone, then the value of $G\square H$ is the sum of the values of G and $\square H$, provided that G and H do not interfere.*

The condition that the games G and H should not interfere means that there may be no overlap between the edges of the squares of G with the edges of the squares of H.

Proof.

$$G\square H \leq G + \square H = G\square + \square H \leq G\square H$$

The first inequality is justified, since splitting a horizontal line can only favor Vertical. The equality is the condition of the theorem. The second inequality is justified since joining two horizontally adjacent squares also can only favor Vertical. $\square$

Following Conway [7] we denote such bridging square as *explosive*. Note that this theorem concerns two fragments connected horizontally. Of course, the theorem is equally valid when the two fragments are vertically connected. Important is that the two fragments are linearly connected via an explosive bridge, i.e., that the bridge has two **opposite** edges in common with the two fragments.

3.2   The Bridge Destroying Theorem for Domineering

When the bridge between two connecting fragments is an explosive square for **both** fragments, then it was shown in [14] that the two fragments need not necessarily be linearly connected, but also may be orthogonally connected via an explosive bridge, i.e., that the bridge has two **adjacent** edges in common with the two fragments. In fact, the explosive square then may act as a bridge between any possible number of fragments from (trivially) zero to the full amount of four. The proof is as follows.

Theorem 2 (Bridge Destroying Theorem). *If for some game $G\square$ its value is equal to that of the game G alone, for some game $\overset{H}{\square}$ its value is equal to that of the game H alone, for some game $\square I$ its value is equal to that of the game I alone, and for some game $\underset{J}{\square}$ its value is equal to that of the game J alone, then the value of $G\overset{H}{\underset{J}{\square}}I$ is the sum of the values of G, H, I, and J, provided that G, H, I, and J do not interfere. Games G, H, I, and J might be empty.*

Proof.

$$G\overset{H}{\underset{J}{\square}}I \le G + \overset{H}{\underset{J}{\square}}I \le G + \overset{H}{\underset{J}{\square}} + I = G + \overset{H}{\square} + I + J$$

$$= G + H + I + J$$

$$= G\square + H + I + J$$

$$= G\square I + H + J \le G\overset{H}{\square}I + J \le G\overset{H}{\underset{J}{\square}}I$$

The first two inequalities are justified, since splitting a horizontal line can only favor Vertical, the first equality is an application of the Bridge Splitting Theorem for vertical connections, the next two equalities are just two conditions of the theorem, the fourth equality is an application of the Bridge Splitting Theorem for horizontal connections, and the last two inequalities are justified, since linking along a vertical line also can only favor Vertical.   □

We note that this proof also includes the cases where any subset of $\{G, H, I, J\}$ are empty games, since the value of the empty square equals the value of an empty game. We further note that for two non-empty games this theorem is covered by the Bridge Splitting Theorem when the games are connected to opposite sides of the bridging square, but that this is not needed for the Bridge Destroying Theorem, i.e., a "corner" connection is also allowed.

3.3   The Bridge Destroying Theorem for Domineering Networks

In this subsection we extend the previous Bridge Destroying Theorem to arbitrarily large networks of fragments. Here, networks are defined as single-component positions with an arbitrary number of fragments connected by bridges to exactly two fragments each (*two-way bridges*). So, we refrain from the requirement that the fragments, apart from the bridging square under consideration, may not interfere. To be more precise, we allow the components to have multiple connections, but every connection should be via an explosive square to exactly two fragments.

Before we give the proof we first, in analogy with Conway's *explosive* squares, define the notion of *superfluous* squares.

Definition 1. A *superfluous* square is a square that does not influence the CGT value of a position. This means that the CGT value of the position is exactly equal with and without that square.

We note that the bridging squares in the formulation of Theorem 2 are superfluous. We also note that it can be the case that several squares are candidates for being superfluous. We then have to choose which square(s) will be denoted as superfluous. As a case in point, the two side squares of a Small-T component (see Fig. 3, top) both satisfy the condition for being superfluous, but once one is denoted as superfluous, the other no longer satisfies the requirement. When candidate superfluous squares do no interfere, e.g., side squares in different Small-T components in some position, they may all be denoted as superfluous. This leads us to the following definition of *sets of independent superfluous squares*.

Definition 2. A set S of superfluous squares in a position is *independent* if the CGT value of the position is exactly equal with and without any subset of S.

In the remainder we consider single-component positions consisting of fragments connected via two-way bridges from an independent set of superfluous squares (further called *aggregates*). Since we only consider two-way bridges, we can define the notion of the *connection graph* of an aggregate, together with some related characteristics.

Definition 3. A *connection graph* of an aggregate is the graph with as nodes the fragments of the aggregate and as edges the bridges.

Definition 4. A *tree-structured aggregate* is an aggregate for which the connection graph is a tree. We also denote this as a *0-cyclic* aggregate.

Definition 5. An *n-cyclic aggregate* is an aggregate that contains at least one bridge such that without the bridge the aggregate is $(n-1)$-cyclic.

Using these definitions we can prove the following theorem.

Theorem 3 (Bridge Destroying Theorem for Networks). *If an aggregate A consists of n fragments $F_1, F_2, \cdots, F_n$ that are connected via bridging squares from an independent set of superfluous squares, the value of the aggregate is equal to the sum of the values of all fragments.*

Proof. We will prove this theorem in two parts. First we will prove the theorem for networks without cycles. Second, using this proof as base case, we will prove the theorem for networks with cycles by induction on the cyclicity of the aggregate.

Base Case: If the aggregate is tree-structured (0-cyclic), then the connection graph of the aggregate is a tree. As a result, there is at least one node with degree 1, i.e., a fragment which is singly connected to the remainder of the aggregate. Removing the superfluous bridging square results in a position consisting of an unconnected fragment and the remainder of the aggregate (if any). Using the Bridge Destroying Theorem the value of the resulting position is the sum of the fragment value and the value of the remainder. Moreover, since the original aggregate was tree-structured, any remainder of the aggregate is also tree-structured, meaning that there is again at least one fragment in the remainder with degree 1, that we can decouple. This process can be continued until all fragments are decoupled, showing that the value of the original aggregate equals the sum of the values of all fragments.

Induction Hypothesis: Assume that for n-cyclic aggregates for some unspecified integer $n \geq 0$ it holds that their values equal the sum of the values of their fragments.

Induction Step: If we have an $(n + 1)$-cyclic aggregate A, it contains by definition a bridging square without which the aggregate is an n-cyclic aggregate A'. Since the value of A' is the sum of the values of the fragments, due to the induction hypothesis, and since adding a superfluous bridging square does not change the value, it follows that the value of A equals the value of A', so also is the sum of the values of the fragments.   □

The complete position A can thus always be built from a tree-structured aggregate with the same fragments (with as value the sum of the values of the fragments according to the base case) by proper addition of bridging squares from an independent set of superfluous squares (induction steps) without changing the value.

4   Domineering Snowflakes

In Subsect. 4.1 we introduce an interesting Domineering position. This position has properties that make the position quite suitable to build large connected polygames of known values. Due to its appearance we dub this position the Domineering *Snowflake*. Next, in Subsect. 4.2 we show how this Snowflake can be used to build larger chains of Snowflakes with known values, based on Theorems 1 and 2. We then show in Subsect. 4.3 that using Theorem 3 we also can build 2-dimensional lattices of Snowflakes with known values, which we call *crystallization of Snowflakes* for obvious reasons.

4.1   The Domineering Snowflake

In [14] we showed that, contrary to expectations [8], there exist many Domineering positions with value *2 and *3. The smallest one, with value *2 is depicted in Fig. 4. We further just call it the Star2 position. Left's first option with three unconnected subgames has value $1* + 0 + -1 = *$; Left's second option with two unconnected subgames has value $1/2 - 1/2 = 0$. The two Right options are their negatives, so also * and 0. Therefore, the proper value of this position is $\{*, 0 | *, 0\} = *2$.

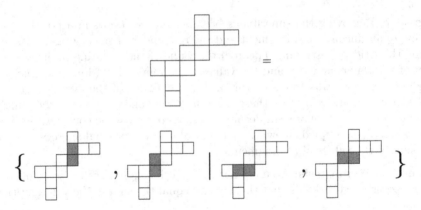

Fig. 4. The Star2 Domineering position, with value *2, and its optimal Left and Right options.

The four squares at the extremes lend themselves for attachment to explosive squares of additional fragments, yielding a position with as value the sum of the component values (applying Theorem 1 four times, twice for a horizontal connection, twice for a vertical connection). When we attach Small-T pieces with value * to the Star2 position, the resulting position is also a nimber, where the value is obtained by appropriate nim additions. So, attaching one Small-T to Star2 gives a position with value *2 + * = *3. Adding a second Small-T then gives *3 + * = *2, a third addition then again gives a *3 position. Finally, adding Small-T's to all extremes gives the position from Fig. 5 with value *2, which we have dubbed the Snowflake position.

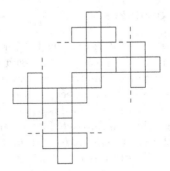

Fig. 5. The Snowflake position, with value *2. The dashed lines show that this position can be considered as the sum of the Star2 position with four Small-T pieces attached.

4.2   Forming Chains of Snowflakes

Since one side square of each Small-T in the Snowflake position is superfluous, we may connect two Snowflakes linearly, applying Theorem 2, to form a chain of two Snowflakes, with value *2 + *2 = 0, see Fig. 6.

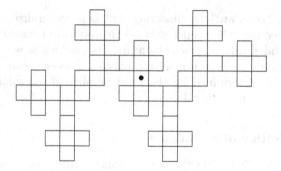

Fig. 6. A chain of two Snowflakes with value 0. The bullet marks the superfluous bridging square.

We can continue adding Snowflakes in the same direction, building arbitrarily long horizontal chains of Snowflakes, where each chain consisting of an even number of Snowflakes has value 0 and each chain with an odd number of Snowflakes has value ∗2. Of course, we symmetrically can build vertical chains of Snowflakes in this way. In fact, we are even able to chain a Snowflake simultaneously in two orthogonal directions, as illustrated in Fig. 7.

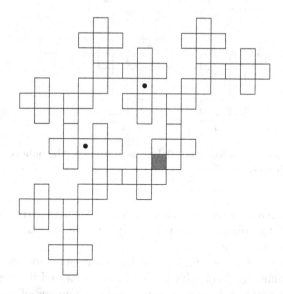

Fig. 7. Three orthogonally connected Snowflakes with value ∗2. The bullets mark the superfluous bridging squares.

The grey cell in this figure denotes a covered square, which means that the fragments here touch, but do not interfere. (This means that the position is not reachable in "standard" Domineering, but is a "generalized" Domineering position [8].) Therefore, the proper value of this position is again *2.

We may extend chains or attach chains likewise, as long as we make sure that chains do not interfere. The proper sum of such a loosely connected network (without cycles) is determined by the total number of Snowflakes being odd (with value *2) or even (with value 0).

4.3   Crystallisation of Snowflakes

If we try to connect four Snowflakes in a square arrangement, we see that we get a cycle introduced. This 2×2 network of Snowflakes is depicted in Fig. 8.

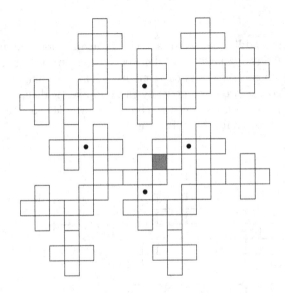

Fig. 8. A square network of four Snowflakes, with value 0. The bullets mark the superfluous bridging squares.

Using Theorem 3 it follows that this connected square matrix still has as CGT value the sum of the values of the four component Snowflakes, i.e., $4 \times *2 = 0$ (a loss for the player to move).

Now this process of connecting Snowflakes in any (linear or orthogonal) direction can be continued to form arbitrary multiple-connected flat networks, where again the whole network has value 0 for an even number of Snowflakes, and value *2 for an odd number. We denote this as *crystallization* of Snowflakes. A snapshot of a part of such a "crystal" is given in Fig. 9, where for clarity all covered cells have been indicated in grey.

Fig. 9. A snapshot of a Snowflake crystal.

5   Conclusions and Future Research

The main contribution of this paper is that we have stated and proved a new, powerful theorem, that indicates when an aggregate consisting of multiple connected networks of fragments has as game-theoretic value the sum of the values of its components.

As an application, we introduced the Snowflake position, with value *2, and showed how from this Snowflake larger chains of Snowflakes can be built, including flat networks (crystals) of Snowflakes, with known values 0 or *2. We note that all game values stated in this paper have extensively been checked using Aaron Siegel's CGSUITE software tool [13].

It is challenging to develop other interesting positions as alternatives for Snowflakes, especially also exhibiting the ability to form arbitrarily large networks with known simple values. Even more challenging is to find a *4 position (and probably accompanying positions with values *5, *6, and *7 by combination with smaller nimbers). It then has to be investigated whether such positions lend themselves also for the kind of crystallization as described for the Snowflakes.

References

1. Albert, M.H., Nowakowski, R.J., Wolfe, D.: Lessons in Play: An Introduction to Combinatorial Game Theory. A K Peters, Wellesley (2007)
2. Breuker, D.M., Uiterwijk, J.W.H.M., van den Herik, H.J.: Solving 8 × 8 Domineering. Theor. Comput. Sci. (Math Games) **230**, 195–206 (2000)
3. Barton, M., Uiterwijk, J.W.H.M.: Combining combinatorial game theory with an α-β solver for Domineering. In: Grootjen, F., Otworowska, M., Kwisthout, J. (eds.) BNAIC 2014: Proceedings of the 26th Benelux Conference on Artificial Intelligence, Radboud University, Nijmegen, pp. 9–16 (2014)
4. Berlekamp, E.R.: Blockbusting and Domineering. J. Combin. Theory (Ser. A) **49**, 67–116 (1988)
5. Berlekamp, E.R., Conway, J.H., Guy, R.K.: Winning Ways for your Mathematical Plays. Academic Press, London (1982). 2nd edn. in four volumes: vol. 1 (2001), vols. 2, 3 (2003), vol. 4 (2004). A K Peters, Wellesley
6. Bullock, N.: Domineering: solving large combinatorial search spaces. ICGA J. **25**, 67–84 (2002)
7. Conway, J.H.: On Numbers and Games. Academic Press, London (1976)
8. Drummond-Cole, G.C.: Positions of value *2 in generalized Domineering and chess. Integers Electr. J. Combin. Number Theory **5**, #G6, 13 (2005)
9. Gardner, M.: Mathematical games. Sci. Am. **230**, 106–108 (1974)
10. van den Herik, H.J., Uiterwijk, J.W.H.M., van Rijswijck, J.: Games solved: now and in the future. Artif. Intell. **134**, 277–311 (2002)
11. Kim, Y.: New values in Domineering. Theor. Comput. Sci. (Math Games) **156**, 263–280 (1996)
12. Knuth, D.E., Moore, R.W.: An analysis of alpha-beta pruning. Artif. Intell. **6**, 293–326 (1975)
13. Siegel, A.N.: Combinatorial game suite: a computer algebra system for research in combinatorial game theory. http://cgsuite.sourceforge.net/
14. Uiterwijk, J.W.H.M., Barton, M.: New results for Domineering from combinatorial game theory endgame databases. Theor. Comput. Sci. **592**, 72–86 (2015)
15. Uiterwijk, J.W.H.M., van den Herik, H.J.: The advantage of the initiative. Inf. Sci. **122**, 43–58 (2000)
16. Wolfe, D.: Snakes in Domineering games. Theor. Comput. Sci. (Math Games) **119**, 323–329 (1993)

First Player's Cannot-Lose Strategies for Cylinder-Infinite-Connect-Four with Widths 2 and 6

Yoshiaki Yamaguchi[1]([⊠]) and Todd W. Neller[2]

[1] The University of Tokyo, Tokyo, Japan
yamaguchi@klee.c.u-tokyo.ac.jp
[2] Gettysburg College, Gettysburg, USA

Abstract. Cylinder-Infinite-Connect-Four is Connect-Four played on a cylindrical square grid board with infinite row height and columns that cycle about its width. In previous work, the first player's cannot-lose strategies have been discovered for all widths except 2 and 6, and the second player's cannot-lose strategies have been discovered with all widths except 6 and 11. In this paper, we show the first player's cannot-lose strategies for widths 2 and 6.

1 Introduction

We begin by introducing the two-player game of Cylinder-Infinite-Connect-Four. We call the first and second players *Black* and *White*, respectively. Cylinder-Infinite-Connect-Four is played on a square grid board that wraps about a semi-infinite cylinder (Fig. 1). Rows extend infinitely upward from the ground, and we number columns of a width w board with indices that cycle rightward from 0 to $w - 1$ and back to 0. Players alternate in dropping disks of their colors to the lowest unoccupied grid cell of each drop column. Thus a game *position*, i.e. a configuration of disks, is unambiguously described as a sequence of column numbers. For clarity, we additionally prefix each column number with the first letter of the player color, so "Bi" or "Wi" means that Black or White, respectively, places a disk in column i.

The object of the game is to be the first player to place four or more of one's own disks in an adjacent line horizontally, vertically, or diagonally. We call such a four-in-a-row a *Connect4*. Because of the cylindrical nature of the board, the Connect4 is further constrained to consist of 4 different disks. Thus, a horizontal Connect4 is not allowed for widths less than 4. If, for a given state and given player strategies, we can show the impossibility of either player ever achieving a Connect4, the value of the game is said to be *draw*.

We call a configuration of disks a position. When a player places a disk, background of the cell is colored gray. We duplicate columns 0 through 2 to the right on wider boards to allow easy inspection of wraparound Connect4 possibilities. Figure 2 shows an example terminal game position after B0W0B2W2B1W3B5. A *threat* is defined as a single grid cell that would complete a Connect4 [3].

© Springer International Publishing Switzerland 2015
A. Plaat et al. (Eds.): ACG 2015, LNCS 9525, pp. 113–121, 2015.
DOI: 10.1007/978-3-319-27992-3_11

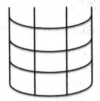

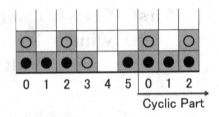

Fig. 1. Board of Cylinder-Infinite-Connect-Four

Fig. 2. Example position

After B0W0B2W2B1, Black has a double threat on the bottom row. Although W3 removes one threat, Black can play the other threat B5 and complete a Connect4.

In previous work [1], the first player's cannot-lose strategies have been discovered for all widths except 2 and 6, and the second player's cannot-lose strategies have been discovered for all widths except 6 and 11. In this paper, we show the first player's cannot-lose strategies for widths 2 and 6.

2   Related Work

In 1988, James Dow Allen proved that Connect-Four played on the standard board with width 7 and height 6 is a first player win [2]; 15 days later, Victor Allis independently proved the same result [4]. Results for Connect-Four games played on finite boards with non-standard heights and/or widths were reported in [9]. Yamaguchi et al. proved that Connect-Four played on a board infinite in height, width, or both, leads to a draw by demonstrating cannot-lose strategies for both players [10]. These cannot-lose strategies are based on paving similar to that used in polyomino achievement games [13–16,18] and 8 (or more) in a row [17].

Other solved connection games include Connect6 for special openings [20], the Hexagonal Polyomino Achievement game for some hexagonal polyominoes [18, 19], Gomoku [11], Renju [12], Qubic [5–7], and Rubik's Cube [8]. Other games with cyclic topology include Cylinder Go [21], Torus Go [21], and TetroSpin [22].

3   First Player's Cannot-Lose Strategy for Cylinder-Infinite-Connect-Four for Width 2

In this section, we show the Black cannot-lose strategy for width 2. First, we define a *follow-up* play as a play in the same column where the opponent just played [4]. Figure 3 shows a Black follow-up play. A *follow-up strategy* is a strategy consisting of follow-up plays.

After Black's first play in Cylinder-Infinite-Connect-Four for width 2, each player has only 2 play choices: follow-up or non-follow-up. Black's cannot-lose strategy is summarized as follows:

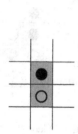

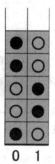

Fig. 3. Black follow-up play   **Fig. 4.** White never plays follow-up

- As long as White plays a non-follow-up strategy, Black alternates between follow-up and non-follow-up plays, starting with follow-up.
- If White plays a follow-up after Black plays a follow-up (Fig. 5), then Black always plays a follow-up strategy thereafter.
- If instead White plays a follow-up after Black plays the initial move or a non-follow-up (Case 2 of Fig. 6), then Black always plays a follow-up except after White plays a non-follow-up after White plays a follow-up at first.

We now consider this strategy in detail. As long as White does not play follow-up, Black's alternating follow-up and non-follow-up play leads to the game sequence B0W1B1W0B1W0B0W1 (Fig. 4) and the resulting pattern permits no Connect4 for either player. If this play pattern continues, the game is a draw. Thus we now need only to consider the ramifications of Black's response to a White follow-up play.

As soon as White makes a follow-up play, there are 2 cases shown in Figs. 5 and 6 which include mirror-symmetric cases as well. These cases capture both of the essentially different situations that may arise in a White non-follow-up play sequence of any length, including 0. When White plays follow-up in White's first and second moves, the "ground" line Figs. 5 and 6 are in Fig. 4 play sequence.

When White plays follow-up at Case 1, Black responds with follow-up thereafter (Fig. 7). If White were to play in column 1, Black's follow-up response would then win, so White must then continue an infinite follow-up sequence in column 0 to draw.

After Case 2, Black plays follow-up (Fig. 8). In Fig. 8, the bold line within figures serves to highlight pieces below that must have been played. If White only plays a follow-up strategy, the game is drawn. However, if White plays non-follow-up in column 0, then Black plays non-follow-up in column 1 and then plays a pure follow-up strategy thereafter. As can be seen in Fig. 8, Black's lowest diagonal Connect4 *undercuts* White's lowest diagonal Connect4 (highlighted with a zig-zag line), so any efforts of White to complete a Connect4 will result in Black completing a Connect4 first.

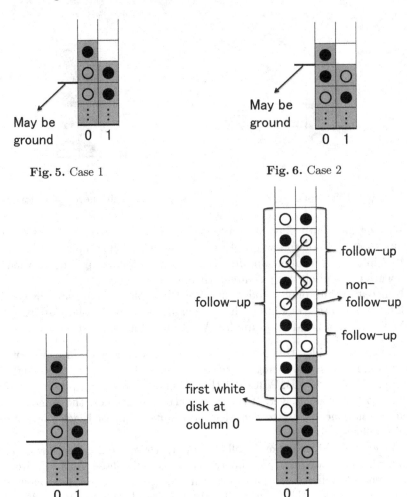

Fig. 5. Case 1   **Fig. 6.** Case 2

Fig. 7. After Case1   **Fig. 8.** After Case2

4   First Player's Cannot-Lose Strategy for Cylinder-Infinite-Connect-Four for Width 6

In this section, we show the Black first player's cannot-lose strategy for width 6 via a branching game-tree case analysis. For each possible line of White play, we show that Black can prevent a White Connect4.

B0W1B2- Fig. 9.
B0W{2 or 3}B0- Fig. 10.
B0W0B2W1B0- Fig. 9.
B0W0B2W3B3W1B0- Fig. 9.
B0W0B2W3B3W{2, 4, or 5}B3- Fig. 11.

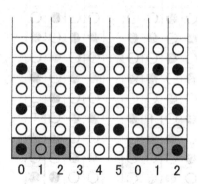

Fig. 9. There is neither Connect4.

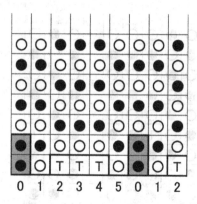

Fig. 10. There is neither Connect4.

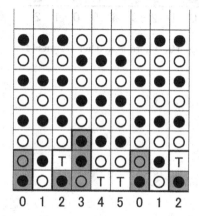

Fig. 11. There is neither Connect4.

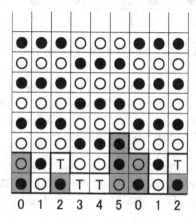

Fig. 12. There is neither Connect4.

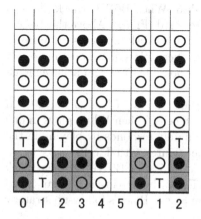

Fig. 13. There is neither Connect4.

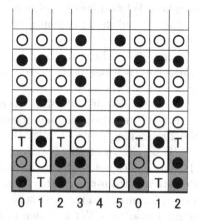

Fig. 14. There is neither Connect4.

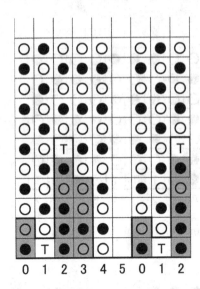

Fig. 15. There is neither Connect4.

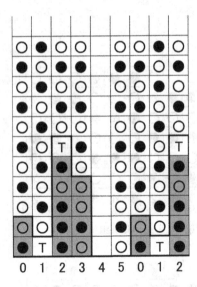

Fig. 16. There is neither Connect4.

B0W0B2W3B3W0B2- Figs. 13 or 14.
B0W0B2W3B3W3B2W{0, 1, or 2}B3- Figs. 13 or 14.
B0W0B2W3B3W3B2W3B2W2B2- Figs. 15 or 16.
B0W0B2W5B5W1B0- Fig. 9.
B0W0B2W5B5W{2, 3, or 4}B5- Fig. 12.
B0W0B2W5B5W0B2- Figs. 17 or 18.
B0W0B2W5B5W5B2W{0, 1, or 2}B5- Figs. 17 or 18.
B0W0B2W5B5W5B2W5B0- Figs. 19 or 20.

We begin our explanation of this case analysis by observing that after Black's play in column 0, we may ignore symmetric board positions and only treat the cases where White plays in columns 0 through 3. When B0W1, Black plays in the column 2 and then plays only follow-up afterward (Fig. 9). Above the highest bold line in all board figures of this section, Black's disk is always on White's disc. Some pieces above the bold line may have been played. All possible 4 × 4 subboards of each board are present so that all possible Connect4 achievements may be visually checked.

After B0W2B0 or B0W3B0, we mark 3 cells with a "T" as in Fig. 10 and note that White's disk occupies one of these 3 T cells. When White plays in one of the 2 remaining empty T cells, Black immediately responds by playing in the other. This pattern of play is repeated in other figures with T cells.

If White does not reply in the column 1, 3, or 5 after B0W0B2, then Black can create a bottom-row double threat (as in Fig. 2) by playing in column 1 and achieve a Connect4 on the next turn in column 3 or 5. Black's follow-up strategy response to B0W0B2W1 is shown in Fig. 9.

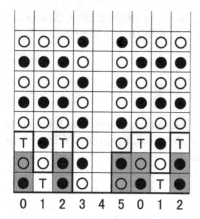

Fig. 17. There is neither Connect4.

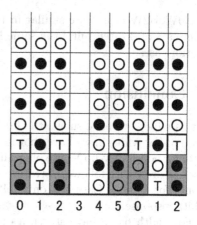

Fig. 18. There is neither Connect4.

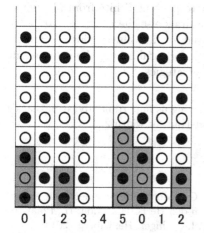

Fig. 19. There is neither Connect4.

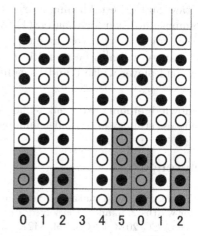

Fig. 20. There is neither Connect4.

After B0W0B2W3B3, White can play in any of the columns 0–5. We show cases B0W0B2W3B3W1B0 and B0W0B2W3B3W{2, 4, or 5}B3 in Figs. 9 and 11, respectively.

After B0W0B2W3B3W0 and B0W0B2W3B3W3, Black plays in column 2 and establishes row 2 threat with playing only follow-up in columns 4 and 5. Because of this 2 threats, White cannot place a disk in both columns 4 and 5. Case B0W0B2W3B3W0B2 is shown in Figs. 13 or 14 where Black plays only follow-up in columns 4 and 5. White can play in any of the columns 0–3 except for columns 4 and 5 after B0W0B2W3B3W3. Cases B0W0B2W3B3W3B2W{0, 1, or 2}B3 are shown in Figs. 13 or 14. Case B0W0B2W3B3W3B2W3B2W2B2 are shown in Figs. 15 or 16 where Black plays only follow-up in columns 4 and 5.

B0W0B2W5 cases are similar in nature to those of B0W0B2W3. Note that columns 3 or 4 often fulfill the same lowest-threat role that columns 4 or 5 did in prior cases.

5 Conclusion

In this paper, we have shown the first player's cannot-lose strategies for widths 2 and 6. For width 2, we have shown that a simple pattern prevents White Connect4 for non-follow-up White play, and that the first follow-up White play allows Black to draw with follow-up strategy after a single non-follow-up response. (This is only for Case 2. Case 1 is only follow-up play. And Black may win by some White response.)

For width 6, we have shown a detailed case analysis where two techniques proved most useful: (1) marking "T" positions to establish a shape for a follow-up draw, and (2) establishing a lowest threat to prevent White from playing in 2 columns and then focusing on (1) to force a draw elsewhere.

We conjecture that the same techniques used for first-player width 6 cannot-lose strategy will also be useful in future work for establishing the same result for second-player width 6 and 11 cannot-lose strategies.

Acknowledgement. We thank I-Chen Wu for giving us important advice on Black's cannot-lose strategy for Cylinder-Infinite-Connect-Four for width 2.

References

1. Yamaguchi, Y., Tanaka, T., Yamaguchi, K.: Cylinder-infinite-connect-four except for widths 2, 6, and 11 is solved: draw. In: van den Herik, H.J., Iida, H., Plaat, A. (eds.) CG 2013. LNCS, vol. 8427, pp. 163–174. Springer, Heidelberg (2014)
2. Allen, J.D.: A note on the computer solution of connect-four. In: Levy, D.N.L., Beal, D.F. (eds.) Heuristic Programming in Artificial Intelligence, The first Computer Olympiad, pp. 134–135. Ellis Horwood, Chinchester (1989)
3. Allen, J.D.: The Complete Book of Connect 4: History, Strategy, Puzzles. Sterling Publishing Co., Inc., New York (2010)
4. Allis, L.V.: A Knowledge-Based Approach to Connect-Four. The game is solved: White wins, Master's thesis, Vrije Universiteit (1988)
5. Mahalko, E.M.: A Possible Win Strategy for the Game of Qubic, Computer Science Master's thesis, Brigham Young University (1976)
6. Patashnik, O.: Qubic: 4 × 4 × 4 Tic-Tac-Toe. Math. Mag. **53**, 202–216 (1980)
7. Allis, L.V.: Searching for Solutions in Games and Artificial Intelligence, Thesis, University of Limburg (1994)
8. God's Number is 20. http://www.cube20.org
9. Tromp, J.: Solving connect-4 on medium board sizes. ACG **31**(2), 110–112 (2008)
10. Yamaguchi, Y., Yamaguchi, K., Tanaka, T., Kaneko, T.: Infinite connect-four is solved: draw. In: van den Herik, H.J., Plaat, A. (eds.) ACG 2011. LNCS, vol. 7168, pp. 208–219. Springer, Heidelberg (2012)

11. Allis, L.V., van den Herik, H.J., Huntjens, M.P.H.: Go-Moku and Threat-Space Search, Report CS 93–02. Faculty of General Sciences, University of Limburg, Department of Computer Science (1993)
12. Wagner, J., Virag, I.: Note solving renju. ICGA J. **24**, 30–35 (2001)
13. Harary, F., Harborth, H.: Achievement and avoidance games with triangular animals. J. Recreational Math. **18**(2), 110–115 (1985–1986)
14. Gardner, M.: Mathematical games. Sci. Amer. **240**, 18–26 (1979)
15. Halupczok, I., Puchta, J.C.S.: Achieving snaky integers. Electron. J. Comb. Number Theor. **7**, G02 (2007)
16. Harary, F.: Is snaky a winner? Geombinatorics **2**, 79–82 (1993)
17. Zetters, T.G.L.: 8 (or More) in a row. Am. Math. Mon. **87**, 575–576 (1980)
18. Bode, J.P., Harborth, H.: Hexagonal polyomino achievement. Discrete Math. **212**, 5–18 (2000)
19. Inagaki, K., Matsuura, A.: Winning strategies for hexagonal polyomino achievement. In: 12th WSEAS International Conference on Applied Mathematics, pp. 252–259 (2007)
20. Wu, I.C.: Relevance-zone-oriented proof search for connect 6. IEEE Trans. Intell. AI Games (SCI) **2**(3), 191–207 (2010)
21. Geselowitz, L.: Freed Go. http://www.leweyg.com/lc/freedgo.html
22. TetroSpin Free APK 1.2. http://m.downloadatoz.com/apps/emre.android.tetrominofree,158809.html

Development of a Program for Playing Progressive Chess

Vito Janko[1]([⊠]) and Matej Guid[2]

[1] Jožef Stefan Institute, Ljubljana, Slovenia
vito.janko@ijs.si
[2] Faculty of Computer and Information Science,
University of Ljubljana, Ljubljana, Slovenia

Abstract. We present the design of a computer program for playing Progressive Chess. In this game, players play progressively longer series of moves rather than just making one move per turn. Our program follows the generally recommended strategy for this game, which consists of three phases: looking for possibilities to checkmate the opponent, playing generally good moves when no checkmate can be found, and preventing checkmates from the opponent. In this paper, we focus on efficiently searching for checkmates, putting to test various heuristics for guiding the search. We also present the findings of self-play experiments between different versions of the program.

1 Introduction

Chess variants comprise a family of strategy board games that are related to, inspired by, or similar to the game of Chess. Progressive Chess is one of the most popular Chess variants [1]: probably hundreds of Progressive Chess tournaments have been held during the past fifty years [2], and several aspects of the game have been investigated and documented [3–6]. In this game, rather than just making one move per turn, players play progressively longer series of moves. White starts with one move, Black plays two consecutive moves, White then plays three moves, and so on.

Rules for Chess apply, with the following exceptions (see more details in [2]):

- Players alternately make a sequence of moves of increasing number.
- A check can be given only on the last move of a turn.
- A player may not expose his own king to check at any time during his turn.
- The king in check must get out of check with the first move of the sequence.
- A player who has no legal move or who runs out of legal moves during his turn is stalemated and the game is drawn.
- En passant capture is admissible on the first move of a turn only.

There are two main varieties of Progressive Chess: Italian Progressive Chess and Scottish Progressive Chess. The former has been examined to a greater extent, and a large database of games (called 'PRBASE') has been assembled. In Italian

© Springer International Publishing Switzerland 2015
A. Plaat et al. (Eds.): ACG 2015, LNCS 9525, pp. 122–134, 2015.
DOI: 10.1007/978-3-319-27992-3_12

Fig. 1. Black to move checkmates in 8 (consecutive) moves.

Progressive Chess, a check may only be given on the last move of a *complete* series of moves. Even if the only way to escape a check is to give check on the first move of the series, then the game is lost by the player in check. In Scottish Progressive Chess, check may be given on any move of a series, but a check also ends the series. It has been shown that the difference very rarely affects the result of the game [7].

The strategy for both players can be summarized as follows. First, look for a checkmate; if none can be found, ensure that the opponent cannot mate next turn. Second, aim to destroy the opponent's most dangerous pieces whilst maximizing the survival chances of your own [2]. Searching for checkmates efficiently – both for the player and for the opponent – is thus an essential task, i.e., the single most important task in this game.

The diagram in Fig. 1 shows an example of a typical challenge in Progressive Chess: to find the sequence of moves that would result in checkmating the opponent. Black checkmates the opponent on the 8th consecutive move (note that White King should not be in check before the last move in the sequence).

Our goal is to develop a strong computer program for playing Progressive Chess. We know of no past attempts to build Progressive Chess playing programs. In the 1990s, a strong Progressive Chess player from Italy, Deumo Polacco, developed ESAU, a program for searching for checkmates in Progressive Chess. According to the program's distributor, AISE (Italian Association of Chess Variants), it was written in Borland Turbo-Basic, and it sometimes required several hours to find a checkmate. To the best of our knowledge, there are neither documented reports about the author's approach, nor whether there were any attempts to extend ESAU to a complete Progressive Chess playing program.

From a game-theoretic perspective, Progressive Chess shares many properties with Chess. It is a finite, sequential, perfect information, deterministic, and zero-sum two-player game. The state-space complexity of a game (defined as the number of game states that can be reached through legal play) is comparable to that of Chess, which has been estimated to be around 10^{46} [8]. However, the per-turn branching factor is extremely large in Progressive Chess, due to the combinatorial possibilities produced by having several steps per turn.

Another chess variant is Arimaa, where "only" four steps per turn are allowed. The branching factor is estimated to be around 16,000 [9]. Up to March 15, 2015, human players prevailed over computers in every annual "Arimaa Challenge" competition. The high branching factor was up to then considered as the main reason why Arimaa is difficult for computer engines [10]. On April 18, 2015, a challenge match between the world's strongest computer program, called SHARP, and a team of three elite human players (Matthew Brown, Jean Daligault, and Lev Ruchka) was completed. Each of the humans played a 3-game match against SHARP. SHARP was given the task to win all three of the individual matches. It did so by 2-1, 2-1, and 3-0, respectively; in total the final score was 7-2 in favor of SHARP (see [11]). Technical details on SHARP are described in [12]. As consequence of this victory in the Arimaa domain, we now expect Progressive Chess to provide a challenging new domain in which we may test new algorithms, ideas, and approaches.

This contribution is organized as follows. In Sect. 2, we describe the design of our Progressive Chess playing program. In Sect. 3, we focus on the specific challenge of searching for checkmates. Experimental design and results of the experiments are presented in Sects. 4 and 5. We then conclude the paper in Sect. 6.[1]

2   Application Description

The graphical user interface of our Progressive Chess playing program is shown in Fig. 2. We implemented the Italian Progressive Chess rules (see Sect. 1 for details). The application provides the following functionalities:

- playing against the computer,
- searching for checkmates,
- watching the computer playing against itself,
- saving games,
- loading and watching saved games.

The user is also allowed to input an arbitrary (but legal) initial position, both for playing and for discovering sequences of moves that lead to a checkmate. The application and related material is available online [13].

2.1   Search Framework

As indicated in the introduction, one of the greatest challenges for AI in this game is its combinatorial complexity. For example, on turn five (White to move has 5 consecutive moves at his[2] disposal) one can play on average around 10^7 different series of moves. Games usually end between turns 5–8, but may lengthen

---

[1] The solution to Fig. 1: Bb4-d6, b6-b5, b5-b4, b4-b3, b3xa2, a2xb1N, Nb1-c3, Bd6-f4.

[2] For brevity, we use 'he' and 'his', whenever 'he or she' and 'his or her' are meant.

Fig. 2. Our Progressive Chess playing program. Black's last turn moves are indicated.

considerably as both players skill increases. Generating and evaluating all possible series for the side to move quickly becomes infeasible as the game progresses. Moreover, searching through all possible responses after each series is even less feasible, rendering conventional algorithms such as minimax or alpha-beta rather useless for successfully playing this game.

Generally speaking, our program is based on heuristic search. However, the search is mainly focused on sequences of moves for the side to move, and to a much lesser extent on considering possible responses by the opponent. In accordance with the aforementioned general strategy of the game, searching for the best series of moves consists of three phases.

Searching for Checkmate. In the first phase, the aim of the search is to discover whether there is a checkmate available. If one is found, the relevant series of moves is executed and the rest of the search is skipped. Checkmates occur rather often, thus finding them efficiently is crucial for successfully playing this game.

Searching for Generally Good Moves. A second search is performed, by trying to improve the position maximally. Usually, the aim of this phase is to eliminate the opponent's most dangerous pieces, and to maximize the survival chances of the own pieces. For example, giving check on the last move of a turn is considered a good tactic, as it effectively reduces the opponent's sequence of moves by one. The king should be given air (e.g., a king on the back rank is often at risk). Pawn promotions are also an important factor to consider. It is often possible to prevent inconvenient opponent's moves by placing the king so that they will give premature check etc. If the allocated time does not allow to search all available series of moves, only a subset of the most promising ones (according to the heuristics) is searched. The series are then ordered based on their heuristic evaluation.

Preventing Checkmate. The previous phase generates a number of sequences and their respective evaluation. It is infeasible to perform a search of all possible opponent replies for each sequence. However, it is advisable to verify whether we are getting mated in the following turn. The most promising sequence of moves is checked for opposing checkmate. In case it is not found, this sequence of moves is then executed. Otherwise the search proceeds with the next-best sequence, and the process then repeats until a safe move is found, or the time runs out. In the latter case, the best sequence according to the heuristic evaluation is chosen. In this phase, again a quick and reliable method for finding checkmates is required.

In Sect. 2.2, we describe the heuristics for finding generally good moves (see the description of the second phase above). In Sect. 3, we describe our approach to searching for checkmates, which is the main focus of this paper.

2.2 Position Heuristics

As described in Sect. 2.1, heuristic evaluation of a particular series of moves is used for finding generally good moves (when checkmate is not available), and taking into account the opponent's replies is often infeasible. Relatively complex heuristics are therefore required. Discovering good heuristics is challenging, as the game is relatively unexplored. Furthermore, a fair bit of what is known to be good in Chess does not apply for Progressive Chess. Defending pieces is an obvious example: while it may be a good idea in orthodox chess, it is often completely useless in Progressive Chess.

We hereby briefly describe the most important heuristics that our program uses for finding sensible sequences of moves (when checkmate is not available):

Material Count. The Shannon value of pieces (Queen = 9, Rook = 5 etc.) hardly apply in Progressive Chess. Bishops are better than Knights in the early stages. In the ending, however, Knights are much better than Bishops because of their ability to reach any square. Pawns are much more dangerous than in the original game, since their promotions often cannot be prevented. Finally, queens are extremely dangerous, because of their huge potential for delivering checkmates. Additional experiments are still required to determine a suitable relative value of the pieces.

King Safety. Kings tend to be safe in the open air, preferably not at the edge of the board. Given the nature of the game it is usually trivial to checkmate a king that is enclosed with its own pieces, so the usual pawn defences or castling are discouraged. Practice showed that king is safest on the second rank; away from opponent pieces, but still safe from back rank mates.

Pawn Placements. Pawns can promote in five moves from their starting position. Stopping them becomes essential as the game progresses. It can be done by block them with pieces, placing pawns in such formation that opposing pawns cannot legally bypass them, or using the King to prevent promotions due to a premature check (but note the Bishop and Knight promotions). Positions where there is no legal promotion from the opponent side are rated higher. It is also favorable to advance pawns, bringing them closer to the promotion square.

Development. Development is a risky proposition, since pieces in the center are more easily captured, and they can often be brought into action from their initial positions rather quickly. Nevertheless, pieces with higher mobility are positively rewarded.

Opening Book. The opening book is upgrading based on results of previous games. The statistics are then used as a part of heuristic evaluation.

Search Extensions. For leaf nodes at the end of the turn it is possible to simulate some opponent replies. Searching only a limited amount of moves may not give an accurate representation of an opponent's best reply, but it gives a general idea. For example, it prevents spending five moves for promoting a pawn that could be taken right on the next move by the opponent.

3   Searching for Checkmate

Searching for checkmates efficiently is the main focus of this paper, i.e., it is our first research question. In this section, we explore various attempts to achieve this goal. It can be considered as a single agent search problem, where the goal is to find a checkmate in a given position. An alternative problem setting would be to find *all* checkmates in the position, or to conclude that a checkmate does not exist, without exploring all possibilities.

The A* algorithm was used for this task. We considered various heuristics for guiding the search. In the experiments, we observed the performance of two different versions of the algorithm (see Sect. 3.1), and of five different heuristics (see Sect. 3.2). Experimental design is described in Sect. 4.

3.1   Algorithm

The task of finding checkmates in Italian Progressive Chess has a particular property – all solutions of a particular problem (position) lie at a fixed depth. Check and checkmate can only be delivered on the last move of the player's turn, so any existing checkmate must be at the depth equal to the turn number. A* uses the distance of the node as an additional term added to the heuristic evaluation, guiding the search towards shorter paths. In positions with a high turn number (where a longer sequence of moves is required) this may not be preferred, as traversing longer variations first is likely to be more promising (as they are the only ones with a solution). One possibility to resolve this problem is to remove the distance term completely, degrading the algorithm into best-first search. An alternative is to weight the distance term according to the known length of the solution. Weight $a/length$ was used for this purpose, where the constant a was set arbitrarily for each individual heuristic. In all versions we

Fig. 3. The values of heuristics listed in Table 1 for this position are as follows. Manhattan: 13, Ghost: 7, Covering: 3, Squares: 8.

acknowledged the symmetry of different move orders and treated them accordingly. In the experiments, we used both versions of the algorithm: *best-first search* and *weighted A**.

3.2   Heuristics

For the purpose of guiding the search towards checkmate positions, we tried an array of different heuristics with different complexities, aiming to find the best trade-off between the speed of evaluation and the reliability of the guidance. This corresponds to the well known search-knowledge tradeoff in game-playing programs [14]. All the heuristics reward maximal value to the checkmate positions. It is particularly important to observe that in such positions, all the squares in the immediate proximity of the opponent's King must be covered, including the King's square itself. This observation served as the basis for the design of the heuristics. They are listed in Table 1.

Figure 3 gives the values of each heuristic from Table 1 for the pieces in the diagram. In *Manhattan*, pawns are not taken into account, resulting in the value of 13 (2+5+6). The value of *Covering* is 3, as the squares a1, a2, b1 are not covered. The *Ghost* heuristic obtains the value of 7 (2+1+2+2): the Rook needs two moves to reach the square immediate to the king, the Bishop needs one, the Knight needs two moves (to reach a1), the Pawn needs one move to promote and one move with the (new) Queen. The value of *Squares* heuristic is 8 (2+3+2+1): the square a1 can be reached in two moves with the Knight, a2 can be reached in three moves with the Knight or Rook, b1 can be reached in two moves with the Rook, b2 can be reached in one move with the Bishop. The player's King is never taken into account in the calculations of heuristic values.

Aside from covering squares around the opponent's King, there are two more useful heuristics that can be combined with the existing ones; we named them *Promotion* and *Pin* (see Table 2).

Table 1. Heuristics for guiding the search for checkmates.

Name	Description
Baseline	Depth-first search without using any heuristic values
Manhattan	The sum of Manhattan distances between pieces and the opponent's King
Ghost	The number of legal moves pieces required to reach the square around the king, if they were moving like "ghosts" (ignoring all the obstacles).
Covering	The number of squares around the king yet to be covered
Squares	The sum of the number of moves that are required for each individual piece to reach every single square around the king.

 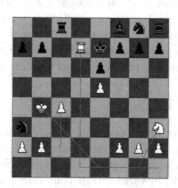

Fig. 4. Left: White to move checkmates in 7 moves. Right: the final position.

A majority of checkmates that occur later in the game include promoting one of the Pawns, getting an extra attacker for delivering checkmate. Rewarding promotions of the Pawns is therefore beneficial.

Another eighth useful heuristic takes advantage of a "self-pin." Figure 4 shows the controversial "Italian mate," which is enthusiastically championed by some but is felt by others to be undesirably artificial [7]. It occurs where the only way to escape a check is to give a check in return, making that move illegal. The position on the left diagram is from a game Boniface – Archer (played in the year 1993), where White played 7 c4, Kd2, Kc3, Kb4, Nf3, Rd1, Rxd7. The final position (diagram on the right) is checkmate according to Italian rules. Our program found an alternative solution (albeit with the same idea), putting the Knight on h3. The moves played were indicated by the program. The solution shows the idea of exploiting the self-pin, moving the King to an appropriate square.

4   Experimental Design

The goal of the experiments was to verify empirically how promising is our approach for finding checkmates in an efficient manner. In particular, (1) which

Table 2. Additional heuristics that can be combined with the existing ones.

Name	Description
Promotion	How far are Pawns to the square of promotion, also rewards extra queens
Pin	How far is the King to the closest square where self-pin could be exploited

of the two search algorithms performs better (*best-first search* or *weighted A**), (2) which is the most promising heuristic to guide the search (*Manhattan*, *Ghost*, *Covering*, or *Squares*), and (3) what is the contribution of the two additional two heuristics (*Promotion* and *Pin*; see Table 2).

A second research question is who has the advantage in Progressive Chess: White or Black (note that this is not so clear as in orthodox chess). *The Classified Encyclopedia of Chess Variants* claims that masters have disagreed on this question, but practice would indicate that White has a definite edge [2].

4.1   Experiment

Two sets of experiments were conducted. First, we observed how quickly do different versions of the program find checkmates on a chosen data set of positions with different solution lengths (see Sect. 4.2). Both average times and success rates within various time constraints were measured. The search was limited to 60 s per position (for each version of the program).

Second, self-play experiments were performed between the programs with the same algorithm (weighted A* with the two additional heuristics) and various other heuristics. The programs played each other in a round-robin fashion. The winning rates were observed for each version of the program, and both for Black and White pieces. In the second phase of the game (see Sect. 2.1), a small random factor influenced the search so the games could be as diverse as possible. Four different, increasingly longer time settings were used in order to verify whether different time constrains affect the performance.

4.2   The Checkmates Data Set

We collected 900 checkmates from real simulated games between programs. In each turn in the range from 4 to 12, there were 100 different checkmates included. The shortest checkmates in Progressive Chess can be made on turn 3, however, they are few and rather trivial. Longer games are rare, and even then there are usually very few pieces left on the board, making the checkmate either trivial or impossible. The above distribution allowed us to observe how the length of the solution affects the search performance.

5 Results

Below we offer three types of results: average times for finding checkmates (5.1), success rates (5.2), and self-play experiments (5.3).

5.1 Average Times for Finding Checkmates

Figure 5 gives the average times for finding checkmates with the best-first search algorithm. It roughly outlines the difficulty of the task: finding checkmates is easier when the solution is short (turns 4–6), more difficult when the solutions are of medium length (turns 7–10), and easier again in the later stage (turns 11–12), as the material on the board dwindles.

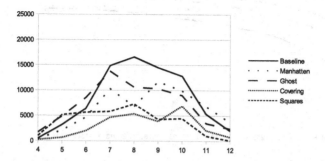

Fig. 5. Average times (in milliseconds) with the best-first search algorithm. The horizontal axis represents the length of the solution.

It is interesting to observe that the baseline heuristic (i.e., depth-first search) even outperforms some other heuristics at turns 4–6 and 11–12, i.e., when the solution is less difficult (note that the problems at higher turns typically contain less material on the chessboard). The Covering heuristic performs best most of the time (up to turn 9), and the Squares heuristic performs best at the later stages.

The average times with the weighted A* algorithm are given in Fig. 6. They are slightly shorter than the ones obtained by the best-first search algorithm up to turn 9. However, the average time increases greatly at the later stages. The main reason is that the heuristics tend to fail to find the solutions in later stages (note that each failed attempt is "penalized" with 60,000 ms, i.e., the time limit for each problem).

However, the performance of the A* algorithm improves drastically when it also uses the two additional heuristics: Promotion and Pin (see Fig. 6). In particular, the Promotion heuristic turns out to be very useful at the later stages in the game.

Overall, the A* algorithm with the two additional heuristics performed best, and Covering heuristic turned out to be the most promising one. This holds, in particular, since most of the games in Progressive Chess finish before turn 10.

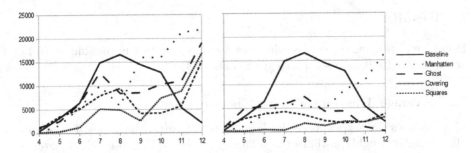

Fig. 6. On the left are average times (in milliseconds) with the A* algorithm, on the right are average times (in miliseconds) with the improved A* algorithm.

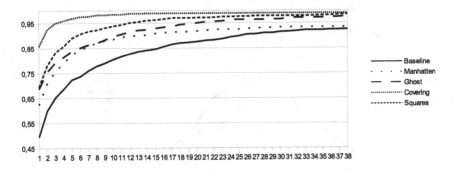

Fig. 7. Percentage of checkmates found in less then x seconds.

5.2   Success Rates

Figure 7 demonstrates how many checkmates were found at any given point of time (in seconds). The Covering heuristic performed clearly best at every cutoff point. It found 80 % (722 out of 900) checkmates in less than a second, and 99 % (891) of the checkmates within the time limit of 60 s. The improved A* algorithm (using the two additional heuristics) was used in this experiment.

5.3   Self-play Experiments

The results of the self-play experiments are given in Fig. 8, showing the number of wins for each heuristic. The Covering heuristic clearly outperformed all the other heuristics, and each heuristic performed better with the black pieces.

There was the total of 31,260 games played, and each program played the same number of games. 20,340 games were played at the time control of 1 s/move, 6,900 at 4 s/move, 2540 at 15 s/move, and 1,480 at 30 s/move. The average length of the games was 8.3 turns ($\sigma = 2.8$). The success rate of white pieces against black pieces was 47.2 % vs. 52.8 %, which suggests that Black has a slight advantage in Progressive Chess. Only 13.7 % of the games ended in a draw.

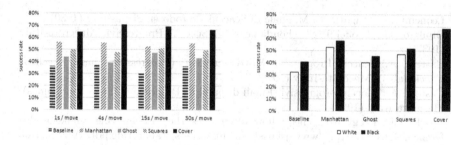

Fig. 8. The success rate for each program (left) and for each piece color (right).

6   Conclusion

The aim of our research is to build a strong computer program for playing and learning Progressive Chess. This chess variant was particularly popular among Italian players in the last two decades of the previous century [2]. By developing a strong computer program, we hope to revive the interest in this game both among human players, who may obtain a strong playing partner and an analysis tool, as well as among computer scientists. In particular, the extremely large branching factor due to the combinatorial explosion of possibilities produced by having several moves per turn makes Progressive Chess both an interesting game and a very challenging environment for testing new algorithms and ideas.

Our program follows the generally recommended strategy for this game, which consists of three phases: (1) looking for possibilities to checkmate the opponent, (2) playing generally good.moves when no checkmate can be found, and (3) preventing checkmates from the opponent. In this paper, we focused on efficiently searching for checkmates, which is considered as the most important task in this game. We introduced various heuristics for guiding the search. The A* algorithm proved to be suitable for the task. In the experiments with (automatically obtained) checkmate-in-N-moves problems, the program found the solutions rather quickly: 80 % within the first second, and 99 % within one minute of search on regular hardware.

A self-play experiment (more than 30,000 games played) between various versions of the program lead to the success rate of 47.2 % vs. 52.8 % in favor of Black. Notably, each version of the program performed better with black pieces.

Our program requires significant further work to achieve the level of the best human players. Particularly in the second phase of the game (which is not directly associated with searching for checkmates) we see much room for improvements, possibly by introducing Monte-Carlo tree search techniques [15]. The question of who has the advantage in Progressive Chess is therefore still open and could be the subject of further investigation.

References

1. Pritchard, D.: Popular Chess Variants. BT Batsford Limited, UK (2000)
2. Pritchard, D.B., Beasley, J.D.: The classified encyclopedia of chess variants. J. Beasley (2007)

3. Leoncini, M., Magari, R.: Manuale di Scacchi Eterodossi, Siena-Italy (1980)
4. Dipilato, G., Leoncini, M.: Fondamenti di Scacchi Progressivi, Macerata-Italy (1987)
5. Castelli, A.: Scacchi Progressivi. Matti Eccellenti (Progressive Chess. Excellent Checkmates), Macerata-Italy (1996)
6. Castelli, A.: Scacchi progressivi. Finali di partita (Progressive Chess. Endgames), Macerata-Italy (1997)
7. Beasley, J.: Progressive chess: how often does the "italian rule" make a difference? (2011). http://www.jsbeasley.co.uk/vchess/italian_rule.pdf. Accessed 06 March 2015
8. Chinchalkar, S.: An upper bound for the number of reachable positions. ICCA J. 19(3), 181–183 (1996)
9. Wu, D.J.: Move ranking and evaluation in the game of Arimaa. Master's thesis, Harvard University, Cambridge, USA (2011)
10. Kozelek, T.: Methods of MCTS and the game Arimaa. Master's thesis, Charles University, Prague, Czech Republic (2010)
11. Lewis, A.: Game over, Arimaa? ICGA J. 38(1), 55–62 (2015)
12. Wu, D.: Designing a winning Arimaa program. ICGA J. 38(1), 19–40 (2015)
13. Janko, V., Guid, M. http://www.ailab.si/progressive-chess/
14. Junghanns, A., Schaeffer, J.: Search versus knowledge in game-playing programs revisited. In: Proceedings of 15th International Joint Conference on Artificial Intelligence, vol. 1, pp. 692–697. Morgan Kaufmann (1999)
15. Coulom, R.: Efficient selectivity and backup operators in Monte-Carlo tree search. In: 5th International Conference on Computers and Games, CG 2006, Turin, Italy, 29–31 May 2006, Revised Papers, pp. 72–83 (2006)

A Comparative Review of Skill Assessment:
Performance, Prediction and Profiling

Guy Haworth[1]([⊠]), Tamal Biswas[1,2], and Ken Regan[2]

[1] The University of Reading, Reading, UK
guy.haworth@bnc.oxon.org
[2] The University at Buffalo (SUNY), Amherst, NY, USA
{tamaltan,regan}@buffalo.edu

Abstract. The assessment of chess players is both an increasingly attractive opportunity and an unfortunate necessity. The chess community needs to limit potential reputational damage by inhibiting cheating and unjustified accusations of cheating: there has been a recent rise in both. A number of counter-intuitive discoveries have been made by benchmarking the intrinsic merit of players' moves: these call for further investigation. Is Capablanca actually, objectively the most accurate World Champion? Has ELO rating inflation not taken place? Stimulated by FIDE/ACP, we revisit the fundamentals of the subject to advance a framework suitable for improved standards of computational experiment and more precise results. Other games and domains look to chess as demonstrator of good practice, including the rating of professionals making high-value decisions under pressure, personnel evaluation by Multichoice Assessment and the organization of crowd-sourcing in citizen science projects. The '3P' themes of performance, prediction and profiling pervade all these domains.

1 Introduction

This position paper is motivated by the recent proliferation of studies and analyses of chess players' skill. A more serious requirement is the need for both scientific rigour and clarity of explanation stemming from the rise in the number of alleged and proven accusations of computer-assisted cheating over the board. FIDE and the Association of Chess Professionals (ACP) are taking n urgent action, involving the third author, to defend the reputation and commercial dimensions of the game [10].

The aim here is to introduce a framework for the discussion and analysis of assessment methods that evaluate the 'intrinsic merit' of the subject's decisions. It clearly separates the goals of metrics for *Performance* and *Prediction*, and quantifies the use of *Profiling*. *Performance* refers to quantitative measures that can be correlated with ELO ratings. *Prediction* means anticipating the distribution of some 100–300 move choices that a player might make in a tournament. Both recognise that strong players will find better moves but anti-cheating tests have to do better than merely identify someone who is 'playing too well'. This is required if analyses of excellent play are to have more statistical significance and more reliably trigger further investigation.

© Springer International Publishing Switzerland 2015
A. Plaat et al. (Eds.): ACG 2015, LNCS 9525, pp. 135–146, 2015.
DOI: 10.1007/978-3-319-27992-3_13

Profiling refers to the use of information about a player's behavior and achievements prior to the period of time relevant to that player's later assessment. Bayesian inference as used by Haworth et al. [1, 3, 5, 6, 15–18] is the natural vehicle for this as it combines an expression of *prior belief* with the modification of that belief in the light of subsequent evidence. Players may also be profiled from a 'cold start' if 'know nothing' priors are adopted. The new FIDE Online Arena with its AceGuard cheating-detection system is notable for its user-profiling which is not done by other statistical methods [9]. Arguably it should have no role in measuring performance but its ability to predict is something that this paper's recommendations will help to study.

1.1   The Chess Engine as Benchmarking Player

All assessment research approaches now use computer-generated move analysis as the benchmark. The reality is that computers are now better and quicker decision makers than humans in the vast majority of chess positions. The top-ranked chess engines are now rated some 300–400 ELO points better than the top human players and no prominent equal-terms match has been played since December 2006. Chess engines deliver their verdicts on the available moves at each nominal ply-depth of their forward-search process. The current top two engines, KOMODO 8, STOCKFISH 6 and others do so via the standard UCI communication protocol [19]. Moves may 'swing up' or 'swing down' as they gain or lose apparent merit at increased depths. This illustrates both the engines' fallibility induced by the finiteness of their vision and the theoretically proven trend to greater precision at higher depths.

However, it is clear that chess engines' centipawn evaluations of positions are not definitive but merely best estimates: engines are fallible agents, even if the best are less fallible than human players. They tend to depart from 0.00 as a decisive result becomes more obvious with increased depth of search, and they vary from engine to engine on the same position [7, 24]. Only in the endgame zone where endgame tables (EGTs) have been computed does an infallible benchmark exist [1, 16–18].

Chess engines operate in one of two modes, *Single-PV* and *Multi-PV*. They focus on what they deem to be the best move in *Single-PV* mode, testing its value rather than also testing 'inferior moves' for comparison. *Multi-PV* guarantees full evaluation to search-depth *sd* of up to *k* 'best' moves as determined by the last round of search. Setting *k* to 50 essentially gives an *sd*-evaluation of every reasonable legal move and many others. *Multi-PV* working requires more time than *Single-PV* working but is required if the full move-context of a move is to be considered.

Alternative-move evaluations are the only input to the benchmarking processes below. No information about the clock-regime, move number, material, relative ELO difference or clock-times is considered, although there is evidence of the error-inducing *zeitnot* effect as players approach move 40 under classic conditions [21].

1.2   A Framework of Requirements for Assessment Methods

The approach here is to re-address the fundamental question 'What are the objectives of player assessment?', to identify the requirements in more detail, and then consider a portfolio of assessment methods in the context of those requirements. All this may be done under our headings of *performance* and *prediction*. As will be seen, all the methods reviewed below measure past performance, some less informed than others. Only a subset of the methods are suitable for reliable prediction.

A more detailed list of requirements is as follows:

1. identifying the current or overall performance of players on some scale,
2. identifying their 'intrinsic skill', i.e., on the basis of their moves, not results,
3. doing so in terms of a 'most likely' scale-point and with c % confidence limits,
4. identifying performance across the years, including the pre-ELO years,
5. ranking players relative to each other on the basis of their intrinsic skill,
6. understanding the stability of methods when subject to small input changes,
7. comparing methods as to the uncertainty budgets associated with their verdicts,
8. using 'robust' methods which are least sensitive to small input changes,
9. improving assessment methods where the opportunity to do so arises,
10. identifying suspected cheating with a suitably high degree of confidence,
11. identifying suspected cheating in real-time in order to trigger further action [11],
12. quashing 'false positive' accusations of over the board cheating,
13. discouraging players from cheating with evidence of good anti-cheating methods,
14. discouraging unfounded accusations of cheating in a similar way, and
15. estimating the probability that player P will play move m in a given position.

The following classification of information illustrates the range of algorithmic sophistication available to a notional punter betting on future moves:

A) Played move and engine-optimal move(s) as evaluated at greatest search-depth,
B) Values of all (reasonable) legal moves as evaluated at greatest search-depth,
C) Values of all (reasonable) moves at all available depths of search,
D) Information about the chess position other than move values, and
E) Information as to a player's tendencies prior to the time of the moves assessed.

Category 'C' highlights the fact that this information has been available but has only recently been recognised as valuable [4]. We argue that 'C' is where the separation of performance and prediction should be focused. The demerit of a superficially attractive move which 'traps' the opponent only becomes visible at the greater depths of search. Heading 'D' includes considerations of pawn structure, attack formations and whether a move is advancing or retreating. Observations on how such factors influence move-choice are made by kibitzers but have not yet been captured by computer algorithm. Time management might be taken into account. Heading 'E' involves considering players' past game and how they might help predict future moves.

2   Useful Notation

The following notation is used in subsequent sections:

- *AP*, the assessed player
- *BP*, the benchmark player against which *AP* is assessed

- *CP*, the cheating player, not only cheating but deceiving an assessment method
- *HP*, the honest player who is not cheating
- *RP*, a *Reference Player*, i.e., a stochastic agent with defined choice-behaviour
- $p_i \equiv$ position i, often implicitly understood to be one of a sequence of positions
- $\{m_j, v_{j,d}\} \equiv$ moves from a position, resulting in values (at depth d) $v_{j,d}$: $v_{j,1} \geq v_{j,2}$ etc.
- $ac_i \equiv$ the *apparent competence* of player *AP* after moving from position p_i.

3   Survey of Assessment Methods

This section gives names to each of the methods known, and lists the methods' absolute and relative advantages ('+'), caveats ('±') and disadvantages ('−'). The first list applies to all methods.

+ chess engines are the only benchmarks which perform consistently across time,
+ the best chess engines are now thought to be better than humans at all tempi,
+ increasing engine ELO decreases move-choice suboptimality by the engine,
+ increasing search-depth increases engine ELO and decreases suboptimality,
+ 'cold start', defined-environment, single-thread running ensures reproducibility,
+ skill-scales may be calibrated in ELO terms using 'Reference ELO e players'
+ skill assessments on such calibrated scales lead to inferred ELO ratings,
± results from different benchmarking engines BP_i may be combined with care,
− *AP's* actual competence varies within games, tournaments and over the years,
− move-choices stem from a plan but are modelled as independent events,
− chess engines are not fully independent, tending to make the same mistakes,
− multithread processing, though attractive, introduces lack of reproducibility,
− there is a probability $p_m > 0$ that cheating player CP will not be detected,
− there is a probability $p_{fp} > 0$ that honest player HP will be accused of cheating.

The eight methods reviewed below are classified under three headings:

- 'Agreement': the observance of agreement between *AP* and *BP*,
- 'Average Difference': the recording of centipawn value 'lost' by *AP*, and
- 'Whole Context': the appreciation of *AP's* move in the full context of options.

3.1   Agreement Between *AP* and *BP*

The methods here are 'MM: Move Matching' and its enhancement 'EV: Equal-value Matching' requiring a little more data as from Multi-PV mode.

MM: Move Matching. Observers of games commonly note whether the human player's choice move matches that of some 'kibitzer-engine' and compute a %-match *MM*. Specific merits (+) and demerits (−) of this method:

+ Engine- and human-moves are easily generated, communicated and compared,
+ the method applies to all moves, even those where engines see 'mate in m',

+ '$MM(AP)$ = 1.00' is a clear 'best possible performance' calibration point,
+ there is no need to scale centipawn values provided by engine BP,
– $MM(AP)$ changes on forced moves when nothing is learned about AP's skill,
– different but equivalent/equi-optimal moves are not regarded as 'matches',
– some engines may randomly choose different equi-optimal moves to top their list,
– cheater CH can easily lower their $MM(CH)$ at minimal cost,
– 'Canals on Mars syndrome': observers are attracted to high-$MM(AP)$ coincidences,
– this method uses the least information of any method.

EV: Equal-value Matching. Disadvantages 2–3 of MM are addressed [2]. Equi-optimal moves are regarded as 'matches', requiring the engines to identify them all:

+ EV is not susceptible to random ordering by chess engine BP whereas MM is,
+ EV results are reproducible whereas MM results from a 'randomising BP' are not,
– EV, unlike MM, requires all equi-optimal moves to be communicated.

3.2 'Average Difference' Methods

AD: Average Difference. Note that we chose AD not AE (Average Error) as the error may come from the benchmarking computer BP rather than from AP [12, 13, 23]:

+ The requisite information is again easily generated, communicated and used,
+ '$AD(AP)$ = 0.00' is a clear 'best possible performance' calibration point,
+ AD, using more information than MM or EV, should give more robust ratings,
– $AD(AP)$ changes on forced moves when nothing is learned about AP's skill,
– $AD(AP)$ changes when AP has only equi-optimal moves to choose from,
– AD only uses information about the best apparent move and AP's choice,
– BP_1 and BP_2 may return different $AD(AP)$, even when choosing the same moves,
– AD cannot be used where the engine gives 'mate in m' rather than an evaluation,
– AD does not scale 'differences' in consideration of the position's absolute value.

AG: Accumulated Gain. This method [8] varies from AD. AP's move is credited with the difference between BP's evaluation of the position before and after the move.

+ AG uses only BP's position evaluation at depth d before and after the played move,
± AG guarantees that the winner will be higher rated than the loser,
– AG conflates the 'horizon effect' with AP's performance,
– AG can give a positive score for a suboptimal move if BP sees a win more clearly,
– AG can penalize an optimal move by the loser as BP sees the win more clearly,
– AG, unlike AD, does not produce a clear mark (0.00) of perfect performance.

Had AG evaluated the position after AP's move at search-depth $d–1$, it would be close to AD. However, it moves BP's horizon on by one ply and therefore credits AP with BP's change of perception one ply later. It does not compare AP's and BP's decisions at the same moment. The concept seems flawed and is not considered further here.

ASD: Average Scaled Difference. The last caveat on the *AD* method anticipates a key finding by Regan [21, 22] that average-difference correlates with the absolute value of the position, see Fig. 1. This may be because (a) humans are only sensitive to the relative values of moves, (b) humans with an advantage tend to avoid the risk associated with the sharpest tactical plans, and/or (c) engines see the win more clearly when the position is relatively decisive already. The case for scaling seems clear.

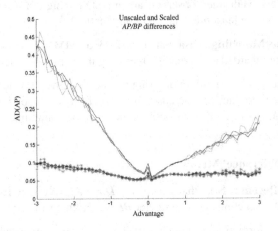

Fig. 1. Unscaled *AP/BP* differences, before and after scaling [21].

If $pv = $ |position value|, AP's 'difference ad relative to *BP's* choice is scaled to be $ad/\ln(1 + pv)$. Regan reports that he now prescales all differences in his industry-scale computations. The recommendation here is that all results produced by the *AD* method should be revisited and scaled in this way.

A detailed study of results from the EV and AD methods [2] also notes the danger of 'false positive' conclusions about suspected cheating over the board. It points to extreme ratings, which any corpus of results will have, which would at first sight be suspicious had they not been achieved before the availability of chess engines at grandmaster level. Table 1 highlights some games analysed with STOCKFISH 3.0.

Table 1. Achievements over the board which would be or are 'false positives' [2]

				Book	Search Moves			
#	Year	White	Black	Res.	depth	Depth	Anal.	CV	Comment
1	1857	Kennicott	Morphy	0-1	29	18	10	—/1.00	Morphy moves 15-24
2	1857	Schulten	Morphy	0-1	8	16	13	—/1.00	Morphy moves 5, 17
3	1866	Morphy	Maurian	1-0	12	18	12	1.00/—	Morphy moves 7, 18
4	1889	Weiss	Burille	1-0	13	20	26	1.00/—	Weiss moves 8-33
5	1965	Carames	Fedorovsky	½-½	18	18		0.85/0.82	Dead drawn, positions 62b-101w
6	1980	Browne	Timman	1-0	33	8	23	1.00/—	Browne moves 18-40
7	2009	Mamedyarov	Kurnosov	0-1	31	var.	6	—/—	too few moves; CV insignificant

3.3 'Whole Context' Analysis: Deepest Evaluations Only

These methods potentially draw on the full context of a move-choice to assess the choice made by *AP*. They deploy a set $SBP \equiv \{BP(c_i)\}$ of stochastic benchmark players of defined competence c_i. As c_i increases, the expected value of BP_i's chosen move increases if this is possible. For these methods:

+ a much fuller use of the move-context is being made,
+ 'apparent competence' does not change if nothing is learned from the move-choice,
+ these methods can easily calculate MM/EV and AD/ASD as byproducts,
− the method potentially requires all moves to be evaluated,
− the method uses the evaluation of the moves at the greatest depth only,
− the number *MultiPV* of 'best moves' considered is a computation parameter,
− the definition of $q_{j,i} \equiv \{Pr[m = m_j \mid BP(c_i)]\}$ requires some domain-specific insight,
− the task of communicating statistical significance is greater than for other methods,
− the results of two *SR* computations cannot easily be combined.

SR: Statistical Regression. This method, deployed by Regan [20–22] identifies the BP_i which best fits the observed play: it is essentially frequentist. The probability of $BP(c_i)$ playing moves m_1-m_k is $p(c_i) \equiv \Pi q_{j,i}$ and c_i is found to maximize $p(c_i)$. The model also generates variances and hence provides *z-scores* for statistical tests employing the MM, EV, and AD/ASD measures.

− the results of two *SR* computations cannot easily be combined.

We report here that SR, carried out to FIDE/ACP guidelines, comes to a negative rather than a positive conclusion on all the games of Table 1, and on the aggregate of Morphy's moves. Given a distribution of MM/EV figures for players of Morphy's standard, the MM/EV figures' z-scores are less than the minimum mark of 2.75 stated [10] as needed to register statistical support for any 'positive' conclusion about cheating likelihood. The Browne-Timman and Mamedyarov-Kurnosov results are less than 0.50. The reason is that the whole-context analysis finds these and Weiss's and Morphy's games to be unusually forcing, so SR gives higher projections than the simpler MM/EV or AD/ASD analyses as [2] would expect. Thus, our category 'B' outclasses 'A' here for the purpose of prediction. This distinction is legislated in [10].

SK: Skilloscopy, the Bayesian approach. Classical probability asks how probable a future event is given a defined scenario. Bayesian analysis asks instead 'What is the probability of each of a set of scenarios given (a) a prior belief in the likelihood of those scenarios and (b) a set of observed events?' An important advantage is that his simple formula can be used iteratively as each new observation arrives.

Skilloscopy is the name given to the assessment of skill by Bayesian Inference [1, 3, 5, 6, 15–18]. It proceeds from initial inherited or presumed probabilities p_i that *AP* 'is' $BP(c_i)$: *AP's* initial presumed apparent competence ac is therefore $\Sigma_i p_i$. Given a move m_j and the probability $q_{j,i} \equiv \{Pr[m = m_j \mid BP(c_i)]\}$, the $\{p_i\}$ are adjusted by the Bayesian formula

$$p_i' \, \alpha \, q_{j,i} \times p_i$$

The $\{p_i'\}$ continue to represent how specifically AP's apparent competence on the c_i-scale is known: AP's apparent competence $ac = \Sigma_i\, p_i'$.

Skilloscopy was first conceived [1, 17, 18] in the context of that part of chess for which endgame tables (EGTs) hold perfect information. These EGTs provided infallible benchmark players BP_c so the above caveats about the fallibility of BP do not apply.

+ SK can combine the results of two independent, compatible computations,

+ SK may evaluate the moves in any order, chronologically or not,

− the choice of $\{BP(c_i)\}$ affects AP_j's rating ac_j after a defined input of evidence,

− AP_j's rating ac_j is meaningful only relative to other ratings ac_j.

3.4 'Whole Context' Analysis: Evaluations at All Depths

The most recent addition to the spectrum of assessment methods [4] is labelled 'SRA' here, being SR but taking move-valuations from all depths of BP's search. It is clear that if a move takes and retains top ranking early in the search, it is more likely to be selected by AP than a move that emerges in the last iteration of the search. Therefore, to ignore shallower-depth evaluations is to ignore valuable information.

Similarly, one can study the way in which such indices as MM/EV and AD/ASD plateau out as search-depth increases. It appears that greater depths are required to get stable MM/EV/ASD ratings for better players. Figure 2 generated from the STOCKFISH v2.31 and v3 data [4] shows this for ASD and also corroborates the contention of Guid and Bratko [12, 13, 23] that even CRAFTY's relatively shallow analysis of world champions suffices to rank them accurately if not to rate them accurately. The sixty players in the 2013 World Blitz championship (WB) had average rating 2611 but showed a competence lower than 2200 at classical time controls.

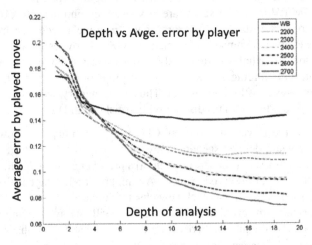

Fig. 2. 'Average Difference' statistics reaching a plateau as BP's search depth increases.

4   The Reference ELO Player

RP_e, a Reference Player with ELO e may be defined by analyzing the moves of a set of players with ELO $e \pm \delta$, e.g., [2690, 2710]. This was done [5, 6, 15], in fact restricting chosen games to those between two such players.[1] The players' ratings in MM/EV, AD/ASD and SR/SK terms may be used to calibrate their respective scales.

Following such calibration, any set of move-choices may be given an Inferred Performance Rating, IPR. That IPR may be placed in the distribution of IPRs by nominally similar players and may be admired or investigated as appropriate (Fig. 3).

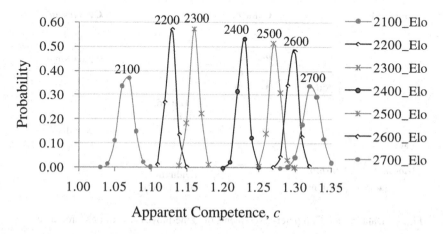

Fig. 3. The set of ELO e Reference Players used by Skilloscopy [5, 6, 15].

5   Standards for a Research Community

The statistical assessment of IPRs requires large amounts of relevant data. The large choice of chess engines, versions, search depths and other computational parameters does not help in combining results by different workers. There is a natural preference to use the best available engines to reduce position-evaluation inaccuracy, and the typical reign of the 'best engine' is usually short.[2]

However, greater interworking in the community may be assisted by:

- the 'separation' of move-analysis, skill-rating and inferred performance rating,
- computational experiments being done in a defined and reproducible way,
- a comprehensive data-model encompassing the computations' results, and
- a robust, accessible repository of results consonant with the data-model about: move analyses, skill-rating exercises and inferences of 'apparent ELO'.

---

[1]   This probably increased the apparent competence ac of RP_e: draws exhibited higher ac.

[2]   Over the last four years, the winners of the TCEC events [24] have been HOUDINI 1.5a, HOUDINI 3, KOMODO 1142, STOCKFISH 170514 and KOMODO 1333.

The reproducibility of computational experiments certainly requires single-thread mode [2], non-learning mode, and the full specification of UCI parameters.[3] Figure 4 is a proposed data-model which may be implemented in Relational or XML databases. The advent of the web-related XML family of standards[4] and the lighter weight 'JSON' Javascript Object Notation have greatly improved the communication and manipulation of hierarchical data.

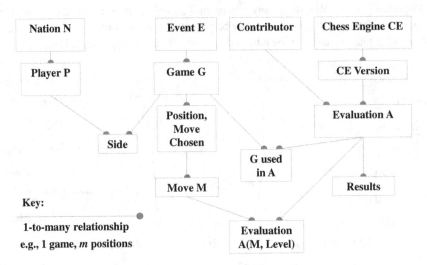

Fig. 4. Data Model: Computer-assessments of game-moves at various search-levels.

6   Summary and View Forward

A number of skill assessment methods have been compared. They vary in their conclusions but differences between workers' computations make definitive comparison difficult at this time.

Greater interworking within the community of those interested in skill assessment is required to quantify the intuitive, widely held but qualitative belief that:

"The more information is used by a method, the better the method is."

Specifically, here, it is believed that MM ⟨ EV ⟨⟨ AD ⟨⟨ ASD ⟨⟨ SR/SK ⟨ SRA.[5] The FIDE/ACP committee certainly regards MM/EV/AD as 'screening methods' but looks to more informed methods for definitive assessments [10].

Therefore the first requirement is to agree on a shared computational approach and on a set of computation subjects in order to quantify the belief above. Agreed tools and data-management interfaces will facilitate progress within the community.

---

[3]   UCI = Universal Chess Interface [19].

[4]   An example of chess-position in XML format is given in [2].

[5]   The notation here is: ⟨ means 'worse than' and ⟨⟨ means 'much worse than'.

Finally, the authors have sought here not only to bring a new coherence to the community of those assessing chess skill but to explore better ways to communicate the subtleties of assessment to the non-specialist and the public. Data supporting this article is freely available and is being evolved [14].

Acknowledgements. In addition to thanking various parties for stimulating discussions on skill assessment, the authors thank David Barnes and Julio Hernandez Castro of the University of Kent for advance and subsequent discussion of their paper [2]

References

1. Andrist, R.B., Haworth, G.M^{c}C.: Deeper model endgame analysis. Theor. Comput. Sci. **349**(2), 158–167 (2005). doi:10.1016/j.tcs.2005.09.044. ISSN 0304-3975

2. Barnes, D.J., Hernandez-Castro, J.: On the limits of engine analysis for cheating detection in chess. Comput. Secur. **48**, 58–73 (2015)

3. Bayes, T.: An essay towards solving a problem in the doctrine of chances. Phil. Trans. Royal Soc. **53**, 370–418 (1763). doi:10.1098/rstl.1763.0053

4. Biswas, T., Regan, K.W.: Quantifying depth and complexity of thinking and knowledge. In: ICAART 2015, the 7th International Conference on Agents and Artificial Intelligence, January 2015, Lisbon, Portugal (2015)

5. Di Fatta, G., Haworth, G.M^{c}C.: Skilloscopy: Bayesian modeling of decision makers' skill. IEEE Trans. Syst. Man Cybern. Syst. **43**(6), 1290–1301 (2013). doi:10.1109/TSMC. 2013.2252893. ISSN 0018-9472

6. Di Fatta, G., Haworth, G.M^{c}C., Regan, K.: Skill rating by bayesian inference. In: IEEE CIDM Symposium on Computational Intelligence and Data Mining (2009)

7. Ferreira, D.R.: The impact of search depth on chess playing strength. ICGA J. **36**(2), 67–90 (2013)

8. Ferreira, D.R.: Determining the strength of chess players based on actual play. ICGA J. **35**(1), 3–19 (2012)

9. FIDE Online Arena with AceGuard: http://www.fide.com/fide/7318-fide-online-arena.html, http://arena.myfide.net/ (2015)

10. FIDE/ACP. Anti-Cheating Guidelines approved by FIDE (2014). http://www.fide.com/ images/-stories/NEWS_2014/FIDE_news/4th_PB_Sochi_Agenda_Minutes/Annex_50.pdf

11. Friedel, F.: Cheating in chess. In: Advances in Computer Games, vol. 9, pp. 327 – 346. Institute for Knowledge and Agent Technology (IKAT), Maastricht (2001)

12. Guid, M., Perez, A., Bratko, I.: How trustworthy is Crafty's analysis of chess champions? ICGA J. **31**(3), 131–144 (2008)

13. Guid, M., Bratko, I.: Computer analysis of world chess champions. ICGA J. **29**(2), 65–73 (2006)

14. Haworth, G.M^{c}C., Biswas, T., Regan, K.W.: This article and related, evolving datasets including pgn and statistics files (2015). http://centaur.reading.ac.uk/39431/

15. Haworth, G.M^{c}C, Regan, K., Di Fatta, G.: Performance and prediction: bayesian modelling of fallible choice in chess. In: van den Herik, H.J., Spronck, P. (eds.) ACG 2009. LNCS, vol. 6048, pp. 99–110. Springer, Heidelberg (2010)

16. Haworth, G.M^{c}C.: Gentlemen, stop your engines! ICGA J. **30**(3), 150–156 (2007)

17. Haworth, G.M^{c}C., Andrist, R.B.: Model endgame analysis. In: van den Herik, H.J., Iida, H., Heinz, E.A. (eds.) Advances in Computer Games: Many Games, Many Challenges, vol. 135(10), pp. 65–79. Kluwer Academic Publishers, Norwell (2004). ISBN 9781402077098

18. Haworth, G.M^{c}C.: Reference fallible endgame play. ICGA J. **26**(2), 81–91 (2003)
19. Huber, R., Meyer-Kahlen, S.: Universal Chess Interface specification (2000). http://wbec-ridderkerk.nl/html/UCIProtocol.html
20. Regan, K.W., Biswas, T.: Psychometric modeling of decision making via game play. In: CIG 2013, the 2013 IEEE Conference on Computational Intelligence in Games, August 2013, Niagara Falls, Canada (2013)
21. Regan, K.W., Macieja, B., Haworth, G.M^{c}C.: Understanding distributions of chess performances. In: van den Herik, H.J., Plaat, A. (eds.) ACG 2011. LNCS, vol. 7168, pp. 230–243. Springer, Heidelberg (2012). doi:10.1007/978-3-642-31866-5_20
22. Regan, K.W., Haworth, G.M^{c}C.: Intrinsic chess ratings. In: AAAI 2011, the 25th AAAI Conference on Artificial Intelligence, 07–11 August 2011, San Francisco, USA, pp. 834–839 (2011). ISBN: 9781-5773-5507-6
23. Riis, S.: Review of "Computer Analysis of World Champions" (2006). http://www.chessbase.com/newsdetail.asp?newsid=3465
24. TCEC: Thoresen's Chess Engine Competition, Seasons 1–7 (2014). http://tcec.chessdom.com/archive.php

Boundary Matching for Interactive Sprouts

Cameron Browne[✉]

Queensland University of Technology, Gardens Point, Brisbane 4000, Australia
c.browne@qut.edu.au

Abstract. The simplicity of the pen-and-paper game Sprouts hides a surprising combinatorial complexity. We describe an optimization called *boundary matching* that accommodates this complexity to allow move generation for Sprouts games of arbitrary size at interactive speeds.

1 Introduction

Sprouts is a combinatorial pen-and-paper game devised by mathematicians Michael S. Paterson and John H. Conway in the 1960s [2], and popularised in a 1967 *Scientific American* article by Martin Gardner [5]. The game is played on a set of n vertices, on which players take turns drawing a path from one vertex to another (or itself) and adding a new vertex along that path, such that $|v_i|$, the *cardinality*[1] of any vertex v_i, never exceeds 3, and no two paths ever touch or intersect (except at vertices). The game is won by the last player able to make a move, so is strictly combinatorial in the mathematical sense.

Figure 1 shows a complete game of $n = 2$ Sprouts between players 1 and 2. Player 2 wins on the fourth move, as player 1 has no moves from this position. A game on n vertices will have at least $2n$ moves and at most $3n - 1$ moves.

It is conjectured that a game on n vertices is a win for player 1 if (n modulo 6) is 0, 1 or 2, and a win for player 2 otherwise [1]. This so called *Sprouts Conjecture* [7] has not been proven yet, but has held for all sizes so far, which include $n = \{1–44, 46–47, 53\}$.[2]

1.1 Motivation

There exist computer Sprouts solvers [1,7] and interactive Sprouts position editors [3], but almost no Sprouts AI players beyond the 3GRAPH player [8] which plays a perfect game up to $n = 8$ vertices, and an iOS player [6] that apparently[3] supports up to $n = 15$ vertices.

This relative lack of Sprouts AI players is something of a mystery. The game is well-known, interesting, looks simple, and has an intuitive topological aspect that just cries out for fingers tracing paths on touch screens – so why are there not more Sprouts apps? We identify the following barriers to implementation:

---

[1] Number of paths incident with vertex v_i.
[2] http://sprouts.tuxfamily.org/wiki/doku.php?id=records.
[3] The app did not work on any device tested.

© Springer International Publishing Switzerland 2015
A. Plaat et al. (Eds.): ACG 2015, LNCS 9525, pp. 147–159, 2015.
DOI: 10.1007/978-3-319-27992-3_14

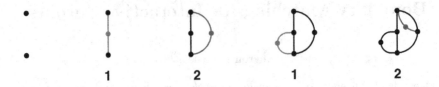

Fig. 1. A game of $n = 2$ Sprouts, won by player 2, who makes the last move.

1. *Complexity:* The game's state space complexity grows at a surprising exponential rate, for even small game sizes [4].[4]
2. *Geometry:* It is non-trivial to synchronise free-form user input curves with an
   underlying algebraic representation of the game.

This paper addresses the first issue by describing an optimization called
boundary matching, which accommodates the game's inherent combinatorial
complexity to reduce move generation time to interactive speeds even for large
game sizes (the issue of geometry will be left for another paper). Section 2 summarizes relevant computational representations for Sprouts, Sect. 3 describes the
boundary matching optimization, Sect. 4 examines the performance of the new
approach, and Sect. 5 discusses its suitability for the task at hand.

2   Representation

Sprouts positions can be represented internally at three levels: *set representation, string representation* and *canonical representation*, as per previous work [1].
We briefly summarise these levels of representation, as necessary background for
describing the Boundary Matching algorithm in Sect. 3.

2.1   Set Representation

A Sprouts *position* is a planar graph obtained according to the rules of the game.
Closed paths resulting from moves divide the position into connected components
called *regions*. Each region contains at least one *boundary* that is a connected
component of the paths made by players and associated *vertices*. Each vertex

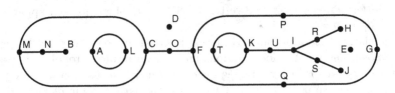

Fig. 2. Example position after ten moves in an $n = 11$ game (from [7]).

---

[4] Denis Mollison's analysis of the $n = 6$ game famously ran to 47 pages [2, p. 602].

may be enclosed within its respective region, or exist on the region border to be shared with adjoining regions.

For example, Fig. 2 shows an example position after ten moves in an $n = 11$ game (from [7]), and Fig. 3 shows the five regions that make up this position.

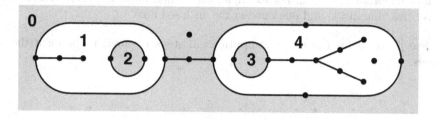

Fig. 3. Region labelling of the position shown in Fig. 2.

A vertex v_i is *alive* if $|v_i| < 3$, otherwise $|v_i| = 3$ and the vertex is *dead*. For example, vertices {A, B, D, E, G, H, J, L, N, O, P, Q, R, S, T, U} in Fig. 2 are alive, whereas vertices {C, F, I, K, M} are dead and play no further part in the game.

2.2   String Representation

We use a slightly simplified version of previous string representations, that we believe is easier to read. Each vertex v is labeled with a unique uppercase character {'A', ..., 'Z'} in these examples. Each boundary B is represented by the list of vertex labels encountered as the boundary is traversed. Each region R is represented by the concatenation of the boundaries it contains, separated by the character ','. A position P is represented by the concatenation of the regions it contains, separated by the character ';'. For example, the position shown in Fig. 2 might be described by the string:

AL;AL,BNMCMN;D,COFPGQFOCM;E,HRISJSIUKTKUIR,FQGP;KT

Subgames: It can be convenient to subdivide positions into independent *subgames*, that describe subsets of regions in which moves cannot possibly affect other subgames. This occurs when all vertices on the border between two such region subsets are dead. Subgames are denoted in the position string by the separating character '/'. For example, the position shown in Fig. 2 can be subdivided as follows:

AL;AL,BNMCMN/D,COFPGQFOCM;E,HRISJSIUKTKUIR,FQGP;KT

Move Types: Before introducing the canonical representation, it is useful to describe the two possible move types. Each move occurs within a region R:

1. *Double-Boundary Moves:* Moves from a vertex v_i on boundary B_m to a vertex v_j on boundary B_n join the two boundaries within R. For example, the first move in Fig. 1 joins two singleton vertices to form a common boundary.
2. *Single-Boundary Moves:* Moves from a vertex v_i on boundary B_m to a vertex v_j on the same boundary (v_i may equal v_j) partition the region R into two new regions R_a and R_b. For example, the second move in Fig. 1 creates two regions, one inside and one outside the enclosed area.

See [7, p. 4] for the exact computational steps required to perform these moves.

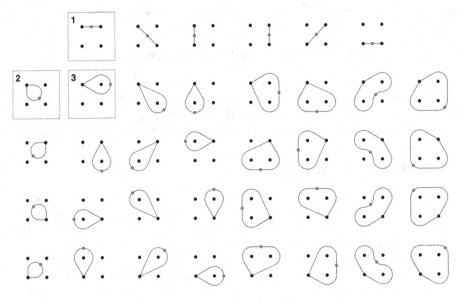

Fig. 4. Opening moves for the $n = 4$ game. Representative invariants are indicated.

Figure 4 shows all possible opening moves for the $n = 4$ game, by way of example. The top row shows the six possible double-boundary moves, and the remaining four rows show the 32 possible single-boundary moves. For any given game size n, the number of opening moves O_n is given by:

$$O_n = \frac{n(n - 1)}{2} + n2^{n-1} \tag{1}$$

If $O_4 = 38$ is a surprising number of opening moves for only four vertices, consider that $O_{10} = 5,165$ opening moves and $O_{20} = 10,485,950$ opening moves(!). The main problem is the exponential growth factor $n2^{n-1}$, due to the number of ways that the region can be partitioned by single-boundary moves.

Invariants: This problem of combinatorial explosion can be addressed by observing that many of the positions in a game of Sprouts are topologically equivalent to others, and can be reduced to a much smaller set of *invariant* forms. For example, each of the 38 positions shown in Fig. 4 are topologically equivalent to one of the three invariant forms indicated, which represent, respectively:

1. all double-boundary moves between one vertex and another;
2. all single-boundary moves that create partitions of $\{0\}$ and $\{3\}$ boundaries;
3. all single-boundary moves that create partitions of $\{1\}$ and $\{2\}$ boundaries.

We therefore only need to consider these three representative cases when evaluating opening moves for the $n = 4$ game. The number of invariant opening forms O_n^+ for a game of size n is given by:[5]

$$O_n^+ = \left\lceil \frac{n}{2} \right\rceil + 1. \tag{2}$$

For comparison, there are six invariant opening forms for the $n = 10$ game from the 5,165 actual opening moves, and only eleven invariant opening forms for the $n = 20$ game from the 10,485,950 actual opening moves. The following section explains how to derive these invariant forms.

2.3   Canonical Representation

Canonical representation involves reducing the string representation of a given position to its canonical (invariant) form. The steps are briefly described below, using the string representation of Fig. 2 as an example:

```
AL;AL,BNMCMN/D,COFPGQFOCM;E,HRISJSIUKTKUIR,FQGP;KT
```

1. Relabel singleton vertices as '0':
   ```
   AL;AL,BNMCMN/0,COFPGQFOCM;0,HRISJSIUKTKUIR,FQGP;KT
   ```
2. Relabel non-singleton vertices that occur exactly once as '1':
   ```
   AL;AL,1NMCMN/0,COFPGQFOCM;0,1RIS1SIUKTKUIR,FQGP;KT
   ```
3. Eliminate vertices that occur three times (i.e. dead vertices):
   ```
   AL;AL,1NN/0,OPGQO;0,1RS1SUTUR,QGP;T
   ```
4. Eliminate boundaries with no remaining vertices and regions with < 2 lives:
   ```
   AL;AL,1NN/0,OPGQO;0,1RS1SUTUR,QGP
   ```
5. Relabel vertices that occur twice in a row along a boundary as '2':
   ```
   AL;AL,12/0,2PGQ;0,1RS1SUTUR,QGP
   ```
6. Relabel vertices that occurred twice but now occur once due to step 4 as '2':
   ```
   AL;AL,12/0,2PGQ;0,1RS1SU2UR,QGP
   ```
7. Relabel vertices that occur twice within a boundary with lower case labels, restarting at 'a' for each boundary:
   ```
   AL;AL,12/0,2PGQ;0,1ab1bc2ca,QGP
   ```

---

[5] Except for $O_1^+ = 1$.

8. Relabel vertices that occur in two different regions with upper case labels, restarting at 'A' for each subgame:
AB;AB,12/0,2ABC;0,1ab1bc2ca,CBA

The resulting string is then processed to find the lexicographically minimum rotation of each boundary within each region, with characters relabeled as appropriate, and sorted in lexicographical order to give the final canonical form:

0,1ab1bc2ca,ABC;0,2ABC/12,AB;AB

Note that the boundaries within a region can be reversed without affecting the result, provided that *all* boundaries within the region are reversed.

It would be prohibitively expensive to perform a true canonicalization[6] that finds the optimal relabeling of uppercase characters for positions in larger games, so we compromise by performing a fast pseudo-canonicalization at the expense of creating some duplicate canonical forms. See [1,7] for details.

3   Boundary Matching

Generating all possible moves for a given position, then reducing these to their canonical forms, is a time consuming process for larger game sizes. Instead, we use a technique called *boundary matching* (BM) to reduce the number of moves that are generated in the first place.

BM works by deriving an invariant form for each boundary *relative to its region*, and moving most of the invariant filtering further up the processing pipeline, before the moves are actually generated. The basic idea is to identify equivalent boundaries within a region, then simply avoid generating duplicate moves from/to boundaries if such moves have already been generated from/to equivalent boundaries in that pass.

3.1   Boundary Equivalence

We describe two regions as *equivalent* if they have the same invariant form, and all live vertices along any member boundary adjoin the same external regions. We describe two boundaries within a region as *equivalent* if they have the same invariant form, and all regions shared by live vertices along each boundary are equivalent.

For example, consider the position shown in Fig. 5, in which region 1 contains the boundaries AK, BC, DF, GH, I and J. Singleton vertices I and J are obviously equivalent boundaries within this region. Boundaries AK, BC and GH are also equivalent within this region, as each adjoin equivalent regions that adjoin back to only region 1. Boundary DF does not have any equivalents, as the region it adjoins (region 3) has a different internal boundary structure.

---

[6] We prefer to use the full term "canonicalization" rather than the abbreviation "canonization"; we do not claim that the algorithms perform miracles.

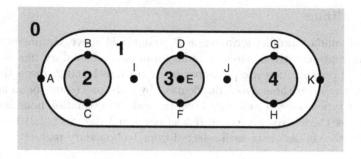

Fig. 5. Within region 1, I and J are equivalent, and AK, BC and GH are equivalent.

The exact steps for calculating region equivalences are not important here, as a singleton optimization (described shortly in Sect. 3.4) simplifies this step.

3.2 Efficient Move Generation

If each equivalent boundary type is assigned an index $t = 1\ldots T$, then a *boundary profile* of each region is provided by the multiset of component boundary type indices. For example, region 1 in Fig. 5 would have the boundary profile $\{1, 1, 2, 2, 2, 3\}$, with $T = 3$ equivalent boundary types.

Double-Boundary Moves: It is only necessary to generate double-boundary moves between unique pairs of equivalent types, where the actual boundaries to be used are selected randomly from the appropriate equivalent subset.

Double-boundary moves would be generated between equivalence types $\{1, 1\}$, $\{1, 2\}$, $\{1, 3\}$, $\{2, 2\}$ and $\{2, 3\}$ in our example. While this provides some improvement, the real savings come from single-boundary moves.

Single-Boundary Moves: It is only necessary to generate single-boundary moves from a single representative of each equivalent boundary type, chosen at random from the appropriate equivalent subset, to itself. Further, it is only necessary to generate partitions of the remaining boundaries within the region, according to the powerset of the region's boundary profile (excluding the source boundary itself).

For our example profile of $\{1, 1, 2, 2, 3\}$, generating single-boundary moves from a representative boundary of equivalence type 1 would generate moves defined by the following partitions: $\{\}$, $\{1\}$, $\{2\}$, $\{3\}$, $\{1,2\}$, $\{1,3\}$, $\{2,2\}$, $\{2,3\}$, $\{1,2,2\}$, $\{1,2,3\}$, $\{1,2,2,2\}$, $\{1,2,2,3\}$, $\{1,2,2,2,3\}$, where the actual boundaries used in each partition are again chosen randomly from the equivalence set with that index.

3.3   Algorithms

Given the boundary profiles for a region R, as outlined above, the algorithms for generating representative double-boundary and single-boundary moves within R using BM optimization are presented in Algorithms 1 and 2, respectively. In each case, the algorithms avoid duplication by only processing boundaries of equivalence type t once in each role as source and/or destination boundary. The actual moves themselves are generated as per usual (see Sect. 2.2) from each pairing of live vertices along each selected from/to boundary pair.

Algorithm 1. Double-Boundary Moves with BM

---

1.   for each boundary type $t_1 = 1 \ldots T$
2.   select boundary b_1 of type t_1 at random
3.   for each live vertex $v_{i=1\ldots I}$ in b_1
4.   for each boundary type $t_2 = t_1 \ldots T$
5.   select boundary b_2 of type t_2 at random, such that $b_1 \neq b_2$
6.   for each live vertex $v_{j=1 \ldots J}$ in b_2
7.   generate double-boundary move from v_i to v_j

Algorithm 2. Single-Boundary Moves with BM

---

1.   for each boundary type $t = 1 \ldots T$
2.   select boundary b of type t at random
3.   for each live vertex $v_{i=1 \ldots I}$ in b
4.   for each live vertex $v_{j=i \ldots I}$ in b
5.   if $i \neq j$ or $|v_i| + |v_j| < 2$
6.   generate single-boundary moves from v_i to v_j (i may equal j)
7.   one partition for each powerset entry (excluding t)

3.4   Singleton Optimization

The approach described above provides significant savings in terms of reducing combinatorial complexity, by avoiding effectively duplicate permutations. However, we can further improve the runtime of the technique by realizing that *singleton* boundaries comprised of a single vertex are: easy to detect; easier to match than other boundary types; the most likely boundary type to match others (on average); and the most common boundary type, at least in the early stages of a game.

It has proven sufficient in our tests to assign all such singleton boundaries to equivalence group 1, and assign each remaining boundary to its own equivalence group without attempting to match it with other boundaries. A typical boundary profile will therefore look something like: {1, 1, 1, 1, 2, 3, 4, ...}, with the

Table 1. Opening move generation without canonicalisation.

n	O_n^+	Without BM		With BM	
		Moves	s	Moves	s
1	1	1	<0.001	1	<0.001
2	2	4	<0.001	3	<0.001
3	3	15	<0.001	4	<0.001
4	3	38	0.002	5	<0.001
5	4	90	0.004	6	<0.001
6	4	207	0.010	7	<0.001
7	5	469	0.029	8	<0.001
8	5	1.052	0.060	9	<0.001
9	6	2,340	0.071	10	<0.001
10	6	5,165	0.067	11	<0.001
11	7	11,319	0.073	12	<0.001
12	7	24,642	0.172	13	<0.001
13	8	53,326	0.413	14	<0.001
14	8	114,772	0.941	15	<0.001
15	9	245,859	2.203	16	<0.001
16	9	524,387	5.261	17	<0.001
17	10	1,114,124	13.408	18	<0.001
18	10	2,358,850	25.110	19	<0.001
19	11	4,798,038	59.962	20	<0.001
20	11	10,473,050	217.342	21	<0.001

occurrences of index 1 reducing as the game progresses and singleton vertices are consumed. This has the same effect as an optimization used by Lemoine and Viennot in their "Glop" combinatorial game solver, which skips all singleton groups except the last in any position during move generation.[7]

The singleton optimization avoids the need for potentially costly boundary matching calculations for more complex cases. It appears that the choice to not match more complex boundaries is compensated by the speed benefit of only matching singleton boundaries, at least for the cases tried so far.

4 Performance

The approaches outlined above were implemented in Java 7 for performance testing. For the boundary matching tests, each boundary profile was ordered by index, converted to a string, and a hash map maintained to store the relevant

---

[7] http://sprouts.tuxfamily.org/wiki/doku.php?id=home.

powerset that contains the single-boundary move partitions for each distinct profile. Powersets are calculated on-the-fly as required whenever a previously unseen profile is encountered. Hence it is useful to seed the table by playing out a number of random games whenever the program changes to a new game size n, so that more commonly needed powersets are pre-generated during initialization rather than during crucial AI thinking time. All timings were made on a single thread of a standard 2 GHz Intel $i7$ machine.

Test #1: Opening Move Generation: The first test concerns the generation of opening moves with and without BM optimization. Table 1 shows the number of opening moves generated for games of size $n = 1$ to 20, and the time required to generate each legal move set, with and without BM optimization.

The computational cost of move generation without BM increases exponentially with n, whereas BM demonstrates linear performance that requires less than a millisecond for legal move generation regardless of n. Note that canonicalization is not applied in either case here; these figures indicate the raw move counts generated by each approach before invariant filtering.

The move counts with BM matching are one less than twice the number of invariant forms in each case, as each partition in the powerset contains a mirror image that is its complement. Unoptimised move generation (i.e. without BM) starts to get too slow for realtime play from around $n = 12$ upwards.

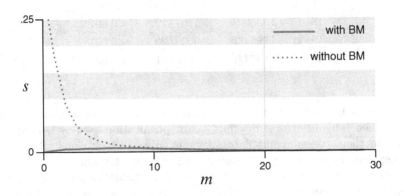

Fig. 6. Convergence of move generation timings over $100 \times n = 10$ games.

Test #2: In-Game Move Generation: The second test concerns the relative performance of the BM optimization over the course of a game. Table 2 shows the average *branching factor* (BF) for each move m over 100 randomly played $n = 10$ games, and timings required to generate the legal move sets with and without BM. Note that full canonicalization is performed in this case, to give accurate branching factors, so timings with and without BM are greater than in Test #1. The ruled line at the $m = 20$ mark indicates the point at which

Table 2. Branching factors and timings over $100 \times n = 10$ games.

m	Samples	BF	Without BM	With BM
			ms	*ms*
0	100	6.00	318.147	0.698
1	100	21.72	183.094	3.141
2	100	25.12	90.098	5.473
3	100	28.30	49.610	5.550
4	100	28.43	30.916	6.632
5	100	25.82	21.304	6.589
6	100	25.74	15.856	7.196
7	100	24.75	11.787	6.968
8	100	22.35	10.233	7.066
9	100	20.98	8.593	6.599
10	100	18.35	6.926	5.704
11	100	16.48	5.524	5.276
12	100	13.89	4.480	4.806
13	100	12.19	3.517	4.115
14	100	10.80	3.246	3.458
15	100	9.15	2.620	2.966
16	100	7.79	2.280	2.465
17	100	7.32	1.986	2.027
18	100	5.85	1.540	1.603
19	100	4.47	1.107	1.273
20	100	3.23	0.817	0.895
21	100	2.37	0.598	0.613
22	100	1.58	0.336	0.365
23	96	0.71	0.133	0.151
24	58	0.26	0.342	0.061
25	15	0.13	0.044	0.017
26	2	0.00	0.010	0.008
27	0	—	—	—
28	0	—	—	—
29	0	—	—	—

games reach the $2n$ mark (recall that all games must last at least $2n$ moves) and typically enter the end game. These results are shown in Fig. 6, in which the solid line indicates timings with BM and the dotted line indicates timings without BM.

Move generation with BM is performed in reasonably constant time through-out the course of each game, despite an increase in branching factor in the early-to-mid game,[8] while move generation with BM takes much longer at the start of each game, dropping quickly until the performance of the two approaches is almost indistinguishable from the mid-game onwards. Similar tests on larger game sizes reveal a similar convergent trend in timings throughout the course of games, although the initial discrepancy becomes much greater as n increases.

5   Discussion

Boundary matching provides significant savings for opening move generation in Sprouts, especially for larger game sizes. For games of $n = 15$ and higher, a single (unoptimized) legal move generation can take much longer than the desired AI thinking time of a few seconds, making BM – or some optimization like it – necessary to achieve realtime response in such cases.

The benefit of BM optimization quickly diminishes until there is little to choose between optimized and unoptimized performance around the mid-game, but it is the early moves that count. An AI player that relies on the lookup of known positions from pre-calculated win/loss tables is more likely to encounter known positions towards the end game, as the game decomposes into simpler sub-games. It is in the opening stages that an AI player, without complete win/loss lookup information for the game size being played, really needs to maximize its lookahead penetration into the game tree.

A caveat with using BM is that the distribution of legal moves produced will not necessarily be the same as that of a random sampling of the search space, which can be a factor if Monte Carlo playouts are involved. For example, the 10,473,050 opening moves of the $n = 20$ game are composed of $n(n-1)/2 = 190$ double-boundary moves and $n2^{n-1} = 10,485,760$ single-boundary moves, in a ratio of 0.000018. However, the BM optimization will only produce 1 (invariant) double-boundary move and 20 (invariant with mirror reflection) single-boundary moves, in a ratio of 0.05. Random sampling can be biased to reflect this inequity (but care must be taken that the noise of single-boundary permutations do not entirely drown out double-boundary moves for larger game sizes) or random moves for playouts can be made directly from the state representation, choosing move type, region and from/to boundaries with the appropriate probabilities.

6   Conclusion

One of the immediate challenges facing the implementation of AI Sprouts players is the problem of move generation at interactive speeds for larger game sizes. The BM optimization offers a solution by allowing fast, near-constant time move generation for arbitrary positions in games of arbitrary size, without any apparent

---

[8] The *early*, *mid* and *end* games could be described as approximately covering moves $1 \ldots n-1, n \ldots 2n-1$ and $2n \ldots 3n-1$, respectively, on average.

drawbacks apart from the potential for random playout bias. Future work might include the investigation of AI search methods, utilizing the BM optimization, for playing the game at arbitrary sizes with minimal corpus knowledge. It would also be worth investigating whether BM performance might be improved through the inclusion of other simple patterns in addition to singletons.

Acknowledgments. This work was funded by a QUT Vice-Chancellor's Research Fellowship as part of the project *Games Without Frontiers*.

References

1. Applegate, D., Jacobson, G., Sleator, D.: Computer Analysis of Sprouts. Technical report CMU-CS-91-144, Carnegie Mellon University Computer Science Technical Report (1991)
2. Berlekamp, E.R., Conway, J.H., Guy, R.K.: Winning Ways for Your Mathematical Plays, vol. 3, 2nd edn. AK Peters, Natick (2001)
3. Department of Mathematics, University of Utah: The Game of Sprouts. http://www.math.utah.edu/~pa/Sprouts
4. Focardi, R., Luccio, F.L.: A modular approach to sprouts. Discrete Appl. Math. **144**(3), 303–319 (2004)
5. Gardner, M.: Mathematical games: of sprouts and brussels sprouts; games with a topological flavour. Sci. Am. **217**(1), 112–115 (1967)
6. Gehrig, D.: Sprouts - A Game of Maths!. https://itunes.apple.com/au/app/spouts-a-game-of-maths!/id426618463?mt=8
7. Lemoine, J., Viennot, S.: Computer Analysis of Sprouts with Nimbers. Technical report, arXiv:1008.2320v1 (2011)
8. Reiß, S.: 3Graph. http://www.reisz.de/3graph_en.htm

Draws, Zugzwangs, and PSPACE-Completeness in the Slither Connection Game

Édouard Bonnet[1], Florian Jamain[2], and Abdallah Saffidine[3]([⊠])

[1] SZTAKI, Hungarian Academy of Sciences, Budapest, Hungary
edouard.bonnet@lamsade.dauphine.fr
[2] LAMSADE, Université Paris-Dauphine, Paris, France
florian.jamain@lamsade.dauphine.fr
[3] CSE, The University of New South Wales, Sydney, Australia
abdallahs@cse.unsw.edu.au

Abstract. Two features set SLITHER apart from other connection games. Previously played stones can be relocated and some stone configurations are forbidden. We show that the interplay of these peculiar mechanics with the standard goal of connecting opposite edges of a board results in a game with a few properties unexpected among connection games, for instance, the existence of mutual Zugzwangs. We also establish that, although there are positions where one player has no legal move, there is no position where both players lack a legal move and that the game cannot end in a draw. From the standpoint of computational complexity, we show that the game is PSPACE-complete, the relocation rule can indeed be tamed so as to simulate a HEX game on a SLITHER board.

1 Introduction

Invented in 2010 by Corey Clark, SLITHER is relatively new connection game with an increasing popularity among online board game players. Unlike HEX and HAVANNAH which are played on a hexagonally-paved board, SLITHER is played on a grid and each player is trying to connect a pair of opposite edges corresponding to their color by constructing connected groups of stones. Whereas moves in most other connection games only involve putting down a new element on the board, moves in SLITHER also allow relocating previously played stones. A second important difference between usual connections games and SLITHER is that some stone configurations are forbidden in SLITHER. Namely, a player is not allowed to play a stone diagonally adjacent to a pre-existing stone of their color unless one of their already placed stones would be mutually adjacent.

The goal of this paper is to study the properties of SLITHER in order to understand better the impact of features such as forbidden configurations and stone relocation on a connection game.

Since its independent inventions in 1942 and 1948 by the Danish poet and mathematician Piet Hein and the American economist and mathematician John Nash, the game of HEX has acquired a special spot in the heart of abstract

© Springer International Publishing Switzerland 2015
A. Plaat et al. (Eds.): ACG 2015, LNCS 9525, pp. 160–176, 2015.
DOI: 10.1007/978-3-319-27992-3_15

game aficionados. Its purity and depth has lead Jack van Rijswijck to conclude his PhD thesis with the following hyperbole [15].

> "HEX has a Platonic existence, independent of human thought. If ever we find an extraterrestrial civilization at all, they will know HEX, without any doubt."

HEX does not only exert a fascination on players, but it is the root of the field of connection games which is being actively explored by game designers and researchers alike [4]. The focuses of the researchers include (1) the design and programming of strong artificial players and solvers [1,6], (2) the computational complexity of determining the winner in arbitrary positions [2,5,11], and (3) theoretical considerations on aspects specific to connection games such a virtual connections and inferior cells [9,13–15]. The two-player game SLITHER that we study in this article should not be confused with the Japanese single-player puzzle SLITHER LINK which has been the object of two other independent papers [16,17].

The paper is organized as follows. After describing the rules of the SLITHER game, we demonstrate via specific game positions that a few properties that typically hold in connection games may actually not hold for SLITHER. The last section establishes that determining the winner in a two-player game of SLITHER is a PSPACE-complete problem.

2   Rules

SLITHER is a two-player game starting on an empty n by n grid (or board). Let us call the players Black (or B) and White (or W). Black and White alternate *moves*. Before stating what a move consists of, and what the winning conditions are, we introduce some useful definitions.

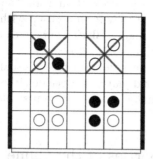

Fig. 1. Some examples of allowed and forbidden configurations. Forbidden configurations are crossed.

As the game proceeds, *squares* of the board can be empty, or contain a black stone, or contain a white stone. We refer to black (resp. white) stones as the

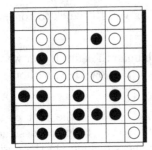

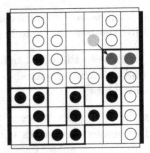

(a) Black to play and win in one move.   (b) The winning move for Black.

Fig. 2. Illustration of a move and a winning group in a 7 by 7 board.

stones of player B (resp. W). We say that two **squares** of the board are *adjacent* if they are in the same row and adjacent columns, or in the same column and adjacent rows. They are *king-adjacent* if a chess king can move from one square to the other, and *diagonally-adjacent* if they are king-adjacent but not adjacent. Two **stones** are *adjacent* (resp. *king-adjacent, diagonally-adjacent*) if they are in adjacent (resp. king-adjacent, diagonally-adjacent) squares. For $P \in \{W, B\}$, let G_P be the graph of which the vertices are the stones of player P placed in the board, and the edges encode the *adjacent* relation. That is, two vertices are linked by an edge if and only if they represent adjacent stones. Like in the game of Go, a *group* for player P is a maximal connected component in G_P.

A *move* for player P consists of an optional relocation of an existing stone of P on a king-adjacent empty grid square, followed by a mandatory placement a stone of P into an empty grid square (see Fig. 2). For a move to be legal, the resulting position may not have two diagonally-adjacent stones of P that do not also have an orthogonally-adjacent stone in common (see Fig. 1). In what follows, we refer to this restrictive rule as the *diagonal rule*.

Black wins if they form a group with at least one stone in the first and in the last column. White wins if they form a group with at least one stone in the first and in the last row. Informally, Black wants to connect left-right and White wants to connect top-bottom (see Fig. 2b).

As in most connection games, a *swap* rule is usually implemented. That is, after the first move, the second player can decide either to play themself a move and the game goes on normally, or to become first player with that very same move.

3   Elementary Properties of the Game

We now present some observations on and properties of SLITHER. Some of the observations made in this section have independently been pointed out earlier in the abstract game community, especially on BoardGameGeek.[1] However, we

---

[1] http://boardgamegeek.com/thread/692652/what-if-there-no-legal-move.

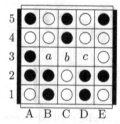
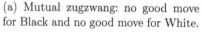
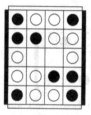

(a) Mutual zugzwang: no good move for Black and no good move for White.

(b) No move allowed for Black, only White has legal moves.

Fig. 3. SLITHER positions with a shortage of moves.

prove for the first time that the standard SLITHER variant cannot end in a draw, thereby settling an open-problem often raised in this community.

The concept of *zugzwang* appears in CHESS and denotes a position in which the current player has no desirable move and would rather pass and have the opponent act. A *mutual zugzwang* is a position in which both players would rather have the opponent play. Although zugzwangs are virtually unheard of in typical connection games, where additional moves can never hurt you, things are different in SLITHER.

Proposition 1. *Zugzwangs and mutual zugzwangs can occur in* SLITHER.

Proof. In Fig. 3a, if it is White (resp. Black) to play, only one move is available, moving stone on C2 (resp. C4) to a and placing a stone on c or equivalently moving to c and placing on a. Then Black (resp. White) wins by placing a stone on C2 (resp. moving stone B5 to C4 and placing a stone on b).

When John Nash discovered/invented the game HEX, one of his motivations was to find a non-trivial game with a non-constructive proof that the first player has a winning strategy in the initial position. To obtain this result, Nash developed a *strategy-stealing argument* which can be summed up as follows [15]. Assume for a contradiction that the second player has a winning strategy σ in the initial position. Then a winning strategy for the first player can be obtained as: start with a random move, then apply σ pretending that the initial move did not occur. If σ ever recommends to play on the location of the initial move, then play another random move and carry on. Given that having an additional random stone on the board cannot hurt the first player, then we have developed a winning strategy for the first player too. Since the second and first player cannot both have a winning strategy at the same time, we may conclude that our hypothesis does not hold. Therefore the initial HEX position is not a second-player win when the swap rule is not used. Nash finishes his proof by showing that there can be no draw in HEX and concludes that the initial position is a first player win.

This strategy-stealing argument can be applied to many other games including TWIXT, HAVANNAH, and games of the CONNECT(m, n, k) family [10]. Since

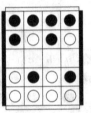

(a) Cylinder board: left and right edges are connected.

(b) Torus board: left and right edges are connected and top and bottom edges are connected.

Fig. 4. Drawn SLITHER positions on non-planar boards.

draws are not ruled-out in these games, the argument only shows that the initial situation is not a second-player win. However, the strategy-stealing argument cannot be applied to SLITHER.

Proposition 2. *Nash's strategy-stealing argument does not apply to* SLITHER.

Proof. Consider the White zugzwang position in Fig. 3a. Had the A4 square not been occupied by a White stone, White would have a winning move: move B4 to *b* and place a stone on *c*. Since there are positions where having one too many stones on the board can make a player lose the game, Nash's strategy-stealing argument does not apply.

Therefore, there is no theoretical indication yet that SLITHER is *not* a second-player win on an empty board. However, in practice it is a huge advantage to play first, so much that if the swap rule is used, it is recommended to swap no matter where the first move is played, including corner locations. The SLITHER-specific intuition behind this practical advice is that the game is dynamic and a player can bring back a stone from a corner towards the center, moving it closer every turn.

Proposition 3. *There exist positions in which a player has no legal move.*

Proof. For instance, Black has no legal move in the position Fig. 3b.

Proposition 4. *Draws are possible when* SLITHER *is played on a cylinder or on a torus.*

Proof. In Fig. 4a (resp. Fig. 4b), if black (resp. black and white) sides are connected, then both players have no legal moves.

That draws are possible for some exotic boards should enhance the accidental aspect of our result on rectangular boards. There are probably no fundamental reasons why the following result is true, and the proof, which we defer to the appendix, consists of a large case analysis on the consequences of forbidding diagonal configurations and the possibility of moving stones.

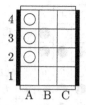

Fig. 5. Ladders.

Theorem 1. *Draws are not possible in* SLITHER *on rectangular boards.*

We end this section, remarking that the connection tactics (for instance ladders) are quite different in SLITHER than in other connection games. For example, in Fig. 5, White is connected to the bottom. Black has to defend on A1, but White can play on C1 and then Black has to play on B1, White plays on C2, Black answers on B2 and White ends the game moving stone A2 to B3 and placing a stone on C3. This last winning move for White is typical in SLITHER and would obviously not be allowed in a game like HEX.

4   Computational Complexity

Here, we show that deciding if one player has a winning strategy from a given position is intractable.

A staple in proofs of PSPACE-hardness for two-player games seems to be GENERALIZED GEOGRAPHY [12]. This problem has been used to show the intractability of games as different as HEX [11], AMAZONS [7], BRIDGE [3], and many more [8]. Our result for SLITHER also relies on GENERALIZED GEOGRAPHY, albeit indirectly since we reduce from HEX.

Theorem 2. *It is* PSPACE-*complete to decide which player has a winning strategy from a given* SLITHER *position.*

Proof. The membership of this problem to PSPACE boils down to noticing that the length of a game is bounded from above by the number of empty squares. Indeed, at each move, one stone is added to the board. Thus, a minimax depth-first search uses a polynomial amount of space.

We now present a reduction from HEX which is known to be PSPACE-complete (see [11]). A hexagonal cell of HEX is encoded by the gadget depicted in Fig. 6. More precisely, an empty cell (resp. a cell containing a black stone, resp. containing a white stone) is transformed into the portion of position of Fig. 6a (resp. Fig. 6b, c). The cell gadgets are glued together and attached to the edges of the board as described in Fig. 7. The empty squares which do not correspond to one of the eight squares designated by letter *a* to *h* in Fig. 6a can be filled with black stones. For convenience, we do not represent those stones.

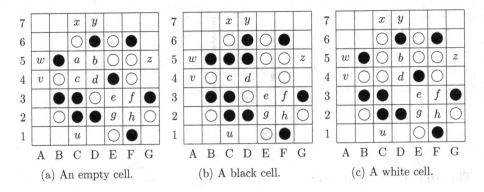

Fig. 6. The cell gadget.

Observation 1. *When a player places a stone in an empty cell gadget (that is, on a square marked with letter a to h), they create a configuration which is forbidden by the diagonal rule. Thus, they should also move one of their stones in the same cell gadget.*

Lemma 1. *In a black cell C (see Fig. 6b), White cannot prevent Black from having a group containing stones in w, y, and u.*

Proof. Black stones on w and y are already in the same group. Because of the diagonal rule, White cannot place a stone in cell C and move a stone in another cell gadget. By Observation 1 and the previous remark, if White moves a stone in cell C, but decides to place a stone in another cell gadget, they can only do so in a white cell, which turns out to be useless. Thus, White might as well place a stone *and* move in cell C. After their move, White should occupy square c; otherwise, Black places a stone on c and thereby connects their group containing u to their group containing w and y.

There are three ways for W to occupy square c: (1) move stone on B4 to c, or (2) move stone on D3 to c, or (3) place a new stone on c. The first option cannot be extended into a legal move. Indeed, the diagonal rule would impose that a stone is placed on d, to connect the two diagonally-adjacent white stones on c and D3. But then white stones on d and $E5$ would form a forbidden configuration. In the second option, White cannot place a stone on D3 nor on d, because of the diagonal rule. And Black's next move would consist of moving the stone on D2 to D3, and placing a stone on d, which connects u to w and y. Finally, in the third option, White is forced to move their stone in D3 to a square other than d. And Black connects in the same manner.

Since the cell gadget is symmetric, the following holds similarly.

Lemma 2. *In a white cell (see Fig. 6c), Black cannot prevent White from having a group containing stones on v, x, and z.*

The following observation is outlined by Fig. 7.

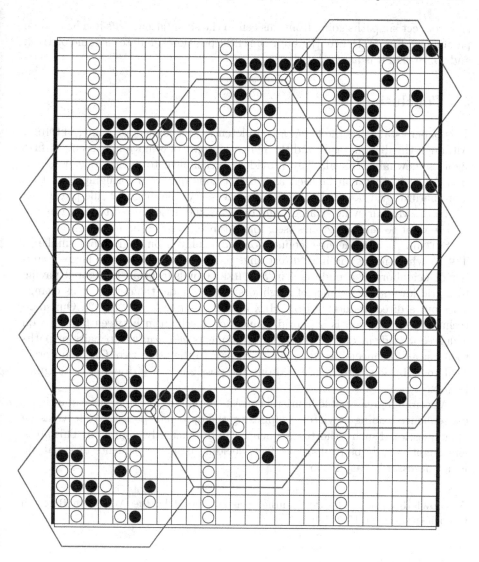

Fig. 7. An empty 3 by 3 HEX board reduced to a SLITHER position.

Observation 2. *When playing in a empty cell gadget, Black cannot do more than connecting u, w, and y, and White cannot do more than connecting v, x, and z.*

From the empty cell gadget, Black can move stone E4 to *b* and place a stone on *a*, resulting in the black cell configuration. By Lemma 1 and Observation 2, it is the optimal play within this cell. Similarly, the optimal play for White in a given empty cell, is to move stone D3 to *c* and place a stone on *a*, yielding the white cell. Thus, having chosen the cell gadget where to play, the optimal move is to connect six paths going from this cell to the six adjacent cells in a hexagonal paving (see Fig. 7). Hence, the built SLITHER position simulates a game of HEX, and so, SLITHER is as hard as HEX.

5 Conclusion

This paper establishes the PSPACE-completeness of a connection game, SLITHER. Although the diagonal rule and the relocation part of a move may seem, at first sight, somewhat artificial, they provide a mechanism to avoid draw possibilities. As such, they can be seen as a direct consequence of bringing HEX to a more natural grid board. Both rules are necessary: the diagonal rule prevents a straightforward drawing strategy, and draws would arise quite often if it was not for the right to relocate stones (see, for instance, Figs. 3a, b, and 7).

SLITHER is not the first connection game played on a grid, but unlike in TWIXT where nodes can have direct links to up to 8 neighbors, the largest number of neighbors of a node is 4 in SLITHER. Going from an hexagonal paving to a grid lowers the number of neighbors of each node, the degree. This change rules out a direct reduction from HEX as was possible for TWIXT [2]. Our main technical contribution is a hybrid connection mechanism between 6 groups on a degree 4 grid: in one move, a player connects *physically* two pairs of paths together and connects *virtually* the third pair to the other two. Our main practical contribution is to have ruled out the possibility of a draw on rectangular boards.

A seemingly reasonable heuristic consists of playing the first move of a shortest sequence leading to victory, supposing the opponent passes. We leave as an open question if such a shortest sequence can be efficiently computed. This problem might already be NP-hard. In fact, deciding if such a sequence exists at all, is not necessarily easy to do.

Acknowledgments. The third author was supported by the Australian Research Council (project DE 150101351).

Appendices

Below we provide four appendices, A, B, C, and D. They provide full proofs of what has stated to be true in the Article.

A Draws are Impossible in SLITHER

Stating that SLITHER does not feature any draw actually corresponds to the following three more elementary statements: each filled SLITHER board has winning groups for at least one player, each filled SLITHER board has winning groups for no more than one player, and each non-filled SLITHER board has at least one legal move for one of the two players. The first two statements can be obtained rather directly from the equivalent ones in HEX, thanks to the diagonal rule. The third statement is much more involved and requires a careful case analysis on non-filled boards.

Lemma 3. *If a* SLITHER *board is filled, then exactly one of the two players wins.*

Proof. Recall that HEX can be played on a rectangular board provided we add a link between each pair of king-adjacent squares along one specified diagonal direction, as in Fig. 8. The forbidden configuration rule ensures that this king-adjacent diagonal connection is respected in SLITHER, although it is indirect. Therefore, any filled $m \times n$ SLITHER board can be mapped onto an equivalent $m \times n$ HEX board such that any pair of SLITHER squares is connected if and only if the corresponding pair of HEX cells is connected. Since any filled HEX board is won by exactly one player, we have the desired SLITHER result.

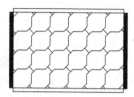

Fig. 8. 6×4 HEX board represented on a rectangular SLITHER board.

Theorem 3. *On a rectangular board, as long as there is at least one empty square, at least one of the two players has a legal move.*

Proof. We adopt the following proof technique.[2] We assume for a contradiction that we have a non-filled position with no legal moves for any players. We start from an empty square and make deduction concerning its surrounding so as to constrain the occupancy of the nearby squares. Each constraint is deduced based on the established occupancies and from the no legal moves assumption or from the diagonal rule. We may perform a split case analysis on squares that are not constrained enough to have a definite status. In each case, however, we finally arrive at a position with a legal move that cannot be prevented by adding any further constraints.

As we add constraints to forbid legal moves from either player, we liberally extend the size of the pattern around the empty square. If any such extension was not possible because we would have reached the limit of the board, then it would not be possible to forbid the desired legal move and our case would be proved. We can therefore disregard the possibility of inadvertently meeting

---

[2] No part of the argument will rely upon the color of the board edge.

an edge of the board as we extend our patterns, at least for the sake of this argument.

In addition to the regular three types of squares, ▢ white stone, ⬤ black stone, and ☐ empty, we add the following ones: ▢ *no constraints yet*, ▢ *cannot hold a white stone*, and ▢ *cannot hold a black stone*.

Consider a rectangular board. If there is at least an empty square on the board, then there is at least an empty square s such that one of the following 4 conditions on the bottom and left neighbors of s is satisfied. Either s is in the bottom-left corner (Fig. 9a), or s is on the bottom edge and its left neighbor is occupied (Fig. 9b), or s has three neighboring stones of different colors (Fig. 9c), or these stones are all of the same color (Fig. 9d). We can assume w.l.o.g. that a majority of the bottom-left neighboring stones are white.

The first two cases are treated in Appendix B, the third case is treated in Appendix C, and the last case in Appendix D. In each case, we arrive to the conclusion that at least one player can move.

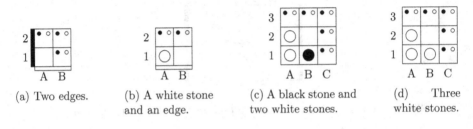

(a) Two edges. (b) A white stone and an edge. (c) A black stone and two white stones. (d)   Three white stones.

Fig. 9. Case analysis for the bottom left surroundings of the empty square.

B   A Square in a Corner or on the Edge of the Board

If there is an empty square in a corner, as depicted in Fig. 9a, then placing a stone on that very square is a legal move for at least one player.

If there is an empty square on an edge, we start from the situation in Fig. 9b and use the following reasoning to constrain the surrounding and obtain Fig. 10. C2 needs a white stone to forbid White's move B1, and C1 cannot be white. A2 needs a black stone to forbid Black's move B1. B2 needs to be empty to forbid White's and Black's move B1. C1 cannot be empty to forbid White's move C1. A3 needs a white stone to forbid White's move A1B1-B2, and B3 cannot be white. C3 needs a black stone to forbid Black's move C1B1-B2, and B3 cannot be black. Similarly, A4 needs a black stone, C4 needs a white stone, and B4 needs to be empty, so as to forbid White's move A1B2-B3 and Black's move C1B2-B3.

But then, C3B2-B1 is a legal move for Black.

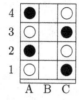

Fig. 10. The case in Fig. 9b with a few deducible constraints filled in.

C   A Square with Two White Stones and a Black Stone

We start from the situation in Fig. 9c and use the following reasoning to constrain the surrounding and obtain Fig. 11. The C2 square cannot contain a white stone, otherwise B2 is a legal move for White. Similarly, B3 cannot contain a black stone, otherwise B2 is a legal move for Black. To forbid White's move B2, there should be a white stone on C1 or on C3 (Fig. 12).

Fig. 11. The case in Fig. 9c with a few deducible constraints filled in.

C.1   Fig. 12a

C2 cannot contain a black stone due to the diagonal rule (since B2 is empty by assumption), and has to be empty. Now, we distinguish two subcases: either B3 is white, or it is empty (Fig. 13).

C.1.1 Fig. 13a. A3 contains a white stone, by the diagonal rule. C3 needs a black stone to forbid Black's move B2. D1 needs a black stone to forbid Black's move B1B2-C2. D3 needs a white stone to forbid White's move C1B2-C2. D2 needs to be empty, by the diagonal rule on stones C1 and C3. E3 needs a black stone to forbid Black's move B1C2-D2.

But, in that situation, D3C2-B2 is a legal move for White.

C.1.2 Fig. 13b. To forbid White's move C2, D3 needs a white stone and C3 and D2 cannot contain a white stone. To forbid White's move D3C2-B2, there should be white stones on E3 and D4 and E4 should not contain a white stone. Now, we distinguish two subcases: either C3 is black, or it is empty (Fig. 14).

(a) Assume C1 is white.

(b) Assume C3 is white.

Fig. 12. Case analysis for Fig. 11, either C1 is white or C3 is white.

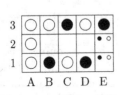

(a) Assume B3 is white.

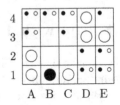

(b) Assume B3 is empty.

Fig. 13. Case analysis for Fig. 12a: B3 is either white or empty.

C.1.2.1 Fig. 14a A3 cannot contain a black stone to forbid Black's move B3. D1 needs a black stone to forbid Black's move C3B2-C2. D2 cannot contain a black stone by the diagonal rule, and has to be empty.

But then, move B1C2-D2 is legal for Black.

C.1.2.2 Fig. 14b B4 needs a white stone to forbid White's move C3. D2 needs a black stone to forbid Black's move C3.

But then, B1C2-C3 is legal for Black.

C.2   Fig. 12b

A3 needs a black stone to forbid Black's move B2. B3 cannot contain a white stone by the diagonal rule (with A2) and has to be empty. Now, we distinguish two subcases: either C2 is black, or it is empty (Fig. 15).

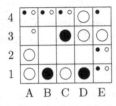

(a) Assume C3 is black.

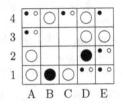

(b) Assume C3 is empty.

Fig. 14. Case analysis for Fig. 13b: C3 is either black or empty.

(a) Assume C2 is black.   (b) Assume C2 is empty.

Fig. 15. Case analysis for Fig. 12b: C2 is either black or empty.

C.2.1 Fig. 15a. To forbid White's move C3B2-B3 and Black's move A3B2-B3, there should be at least one white stone and one black stone in {A4, C4}. So, we distinguish further between {A4 black, C4 white} and {A4 white, C4 black} (Fig. 16).

(a) Assume A4 is black and C4 is white.   (b) Assume A4 is white and C4 is black.

Fig. 16. Case analysis for Fig. 15a: the contents of A4 and C4 is white.

C.2.1.1 Fig. 16a. D2 needs a black stone to forbid Black's move C2B2-B3, and C0 needs a black stone to forbid Black's move C1B2-B3.

But then, B1B2-B3 is a legal move for Black.

C.2.1.2 Fig. 16b. By the diagonal rule, square B4 has to be empty. C5 (as well as D4) needs a black stone to forbid Black's move C4B3-B2. Symmetrically, A5 needs a white stone to forbid White's move A4B3-B2.

But then, C3B2-B4 is a legal move for White.

C.2.2 Fig. 15b. To forbid White's move C2, there should be a white stone on D1 but no white stones on C1 nor D2. We distinguish two subcases: C1 contains a black stone, or it is empty (Fig. 17).

C.2.2.1 Fig. 17a. D3 needs a black stone to forbid Black's move C2 (since B3 is empty). Then, by the diagonal rule, square D2 can only be empty. E3 needs a white stone to forbid White's move C3D2-C2. E1 needs a black stone to forbid Black's move D3C2-D2.

But then, D1C2-B2 is a legal move for White.

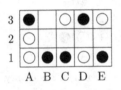

(a) Assume C1 is black.

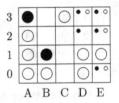

(b) Assume C1 is empty.

Fig. 17. Case analysis for Fig. 15b: C1 is either black or empty.

C.2.2.2 Fig. 17b. The only way to forbid White's move D1C2-B2 is to add two white stones on D0 and on E1. To forbid White's move C1, there should be a white stone on B0 (and a white stone on A0, by the diagonal rule), and no white stones on C0. C0 cannot contain a black stone because of the diagonal rule.

But then, C0 is a legal move for White.

D   A Square with Three White Stones

We start from the situation in Fig. 9d. To forbid White's move B2, there should be a white stone on C3, but no white stones on B3 nor C2. To forbid Black's move B2, there should be a black stone on C1 or A3, say C1 w.l.o.g. (see Fig. 18).

Fig. 18. The case in Fig. 9d with a few deducible constraints filled in.

Therefore, B3 and C2 are empty or contain a black stone. They cannot both contain a black stone since B2 is empty. We thus distinguish three cases: B3 and C2 are empty, B3 contains a black stone, and C2 contains a black stone (Fig. 19).

D.1   Fig. 19a

D3 needs a black stone to forbid Black's move C2. D1 needs a white stone to forbid White's move C3B2-C2. E3 needs a white stone to forbid White's move C3B2-D2. E1 needs a black stone to forbid Black's move D3C2-D2.

But then, D1C2-B2 is legal for White.

D.2   Fig. 19b

A3 needs a black stone to forbid Black's move B2. This case is equivalent to the case of Fig. 15a under color and spatial symmetry.

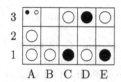

(a) Assume B3 and C2 are empty.

(b) Assume B3 is empty and C2 is black.

(c) Assume B3 is black and C2 is empty.

Fig. 19. Case analysis for Fig. 18: the contents of B3 and C2.

D.3   Fig. 19c

Let us consider cases for the contents of A3. If A3 contains a black stone, then we obtain a position equivalent to Fig. 15a under color and spatial symmetry.

If A3 is empty or white, then a similar proof to Fig. 19a still holds. Indeed, D3 needs a black stone to forbid Black's move B3B2-C2 and D1 needs a white stone to forbid White's move C3B2-C2. The same way, E3 needs a white stone to forbid White's move C3B2-D2, and then E1 needs a black stone to forbid Black's move C1B2-D2.

But then, D1C2-B2 is legal for White.

References

1. Arneson, B., Hayward, R.B., Henderson, P.: Monte Carlo tree search in Hex. IEEE Trans. Comput. Intell. AI Games **2**(4), 251–258 (2010)
2. Bonnet, É., Jamain, F., Saffidine, A.: Havannah and TwixT are PSPACE-complete. In: van den Herik, H.J., Iida, H., Plaat, A. (eds.) CG 2013. LNCS, vol. 8427, pp. 175–186. Springer, Heidelberg (2014)
3. Bonnet, É., Jamain, F., Saffidine, A.: On the complexity of trick-taking card games. In: Rossi, F. (ed.) 23rd International Joint Conference on Artificial Intelligence (IJCAI), Beijing, China, August 2013, pp. 482–488. AAAI Press (2013)
4. Browne, C.: Connection Games: Variations on a Theme. A K Peters, Massachusetts (2005)
5. Even, S., Tarjan, R.E.: A combinatorial problem which is complete in polynomial space. J. ACM (JACM) **23**(4), 710–719 (1976)
6. Ewalds, T.: Playing and solving Havannah. Master's thesis, University of Alberta (2012)
7. Furtak, T., Kiyomi, M., Uno, T., Buro, M.: Generalized Amazons is PSPACE-complete. In: Kaelbling, L.P., Saffiotti, A. (eds.) 19th International Joint Conference on Artificial Intelligence (IJCAI), pp. 132–137 (2005)
8. Hearn, R.A., Demaine, E.D.: Games, Puzzles, and Computation. A K Peters, USA (2009)
9. Henderson, P.T.: Playing and solving the game of Hex. Ph.D. thesis, University of Alberta, August 2010
10. Hsieh, M.Y., Tsai, S.-C.: On the fairness and complexity of generalized-in-a-row games. Theor. Comput. Sci. **385**(1–3), 88–100 (2007)
11. Reisch, S.: Hex ist PSPACE-vollständig. Acta Informatica **15**(2), 167–191 (1981)

12. Schaefer, T.J.: On the complexity of some two-person perfect-information games. J. Comput. Syst. Sci. **16**(2), 185–225 (1978)
13. Steane, A.M.: Threat, support and dead edges in the Shannon game, October 2012
14. Steane, A.M.: Minimal and irreducible links in the Shannon game, January 2013
15. van Rijswijck, J.: Set colouring games. Ph.D. thesis, University of Alberta, October 2006
16. Yato, T.: On the NP-completeness of the Slither link puzzle. In: Notes of the 74th Meeting of IPSJ SIG ALgorithms, pp. 25–32 (2000)
17. Yoshinaka, R., Saitoh, T., Kawahara, J., Tsuruma, K., Iwashita, H., Minato, S.-I.: Finding all solutions and instances of Numberlink and Slitherlink by ZDDs. Algorithms **5**(2), 176–213 (2012)

Constructing Pin Endgame Databases
for the Backgammon Variant Plakoto

Nikolaos Papahristou[✉] and Ioannis Refanidis

University of Macedonia, Thessaloniki, Greece
nikpapa@gmail.com, yrefanid@uom.gr

Abstract. PALAMEDES is an ongoing project for building expert playing bots that can play backgammon variants. Until recently the position evaluation relied only on self-trained neural networks. This paper describes the first attempt to augment PALAMEDES by constructing databases for certain endgame positions for the backgammon variant of Plakoto. The result is 5 databases containing 12,480,720 records in total; they can calculate accurately the best move for roughly 3.4×10^{15} positions. To the best of our knowledge, this is the first time that an endgame database is created for this game.

1 Introduction

Computer game programs are using endgame databases to great effect, especially in board games. Examples of complex games benefiting from such databases are chess [6], Chinese chess [4], checkers [10], awari [8], Kriegspiel [3], and nine-men morris [5], to name a few. Moreover endgame databases are catalytic in every attempt to solve a game, as can be seen in solved games like checkers [11], nine-men morris [5] and more recently heads-up limit texas holdem poker [2].

An endgame database usually contains precomputed game-theoretical values (or near perfect heuristics) for each position record. The game playing program can use this database by searching the records when an endgame position contained in the database is reached by the AI search. The benefits for the program are multiple: (1) the value retrieved from the database is more accurate than the program's evaluation function; (2) the retrieval of the database value is typically faster than the evaluation function execution speed; (3) there is no need to search any further down the tree.

The endgame databases can also provide a powerful analytical tool for the game professional and for understanding the game in general. A prominent example is chess, where positions which humans had analyzed as draws were proven winnable and vice-versa. Also the database constructed when heads-up limit texas holdem poker was solved [2] offered insights that contradicted some human beliefs about the best play in this game.

Backgammon programs also make use of endgame databases. These usually cover the positions where both players have their checkers in the bearoff quadrant (also known as *bearoff* databases). In the two-sided version, these databases offer the game-theoretic

© Springer International Publishing Switzerland 2015
A. Plaat et al. (Eds.): ACG 2015, LNCS 9525, pp. 177–184, 2015.
DOI: 10.1007/978-3-319-27992-3_16

value of the position whereas in the one-sided version, the goal being to minimize the average number of rolls to bearoff, a distribution of the expected number of rolls to bearoff. The one-sided version is much smaller than the two-sided version but it is not as accurate with respect to finding the best move.

Our main focus is on several backgammon variants that are not yet sufficiently examined by computer games research. The aforementioned bearoff endgame databases can be used in many of the variants that we are interested in (Portes, Plakoto, Fevga, Narde), since the bearoff positions of backgammon can occur in all of these games as well. For this purpose, PALAMEDES already contains a two-sided bearoff database that was constructed using similar techniques used by other programs. This database gives the game theoretical value of all bearoff positions when the doubling cube is not used and is 5.48 GB in size.

This paper describes our efforts of our first attempt to construct endgame databases for positions only seen in the Plakoto variant. To the best of our knowledge, this is the first time this kind of endgame database is constructed. We believe this is the first step towards constructing bigger and better endgame databases in the future. Section 2 presents the rules of the Plakoto variant and introduces the position types that are stored in the databases. Section 3 describes the algorithm used to compute the databases. Section 4 discusses several issues including perspectives on building larger databases. Finally, Sect. 5 concludes and gives five avenues for future work.

2   The Rules of the Plakoto Variant

This section presents the rules of the Plakoto variant (Sect. 2.1) and analyzes the positions that we are interested in storing in the databases (Sect. 2.2). In Sect. 2.3 the number of endgame positions will be discussed.

2.1   The Plakoto Variant

The general rules of the game are the same as the regular backgammon apart from the procedure of hitting. Players start the game with fifteen checkers placed in opposing corners (Fig. 1a) and move around the board in opposite directions until they reach the home board which is located opposite from the starting area. Players can win a single or double game as usual but there is no triple win as in backgammon. Finally, the doubling cube is not typically used.

The key feature of the Plakoto variant is the ability to pin hostile checkers, so as to prevent their movement. When a checker of a player is alone in a point, the opponent can move a checker of his own in this point thus pinning (or trapping) the opponent's checker. This point counts then as a "*made point*" as in regular backgammon, which means that the pinning player can move checkers in this point while the pinned player cannot. The pinned checker is allowed to move normally only when all opponent pinning checkers have left the point (unpinning). A more detailed explanation of the Plakoto rules can be found in http://ai.uom.gr/nikpapa/Palamedes/manual/#Plakoto.

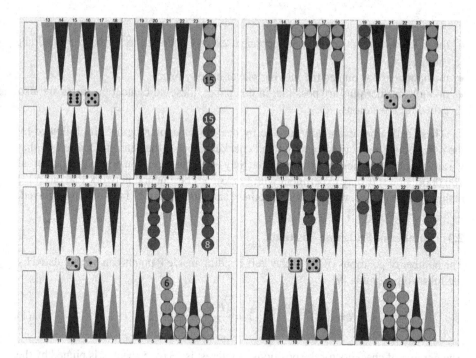

Fig. 1. Various Plakoto positions: (a) Upper left: Starting position. Red player starts at point 1 and bears off at point 24, while green player starts at point 24 and bears off at point 1, (b) Upper right: Typical middle-game position, (c) Lower Left: Endgame position where both players have pins in their bearoff quadrant, and (d) Lower right: Both players have pins in their bearoff quadrants and some checkers in the previous quadrant (Color figure online).

2.2 Endgames with Pins

Strategically thinking, pinning is the most important characteristic of the Plakoto variant for the following three reasons.

1. A "made point" can be constructed with only one checker instead of the usual two which makes primes and other formations easier.
2. Players can nullify bad luck when they roll small rolls and/or the opponent rolls big rolls. This is true because running to the bearoff phase is unimportant when one or more checkers are trapped.
3. The side that has pinned without getting pinned usually gets a few rolls ahead in the bearoff race. The further ahead the pin is, the bigger the advantage.

A typical occurrence in a Plakoto game is for both players to have pinned each other. Then the best strategy usually is to try to maintain the pin(s) for as long as possible trying to make the opponent unpin his own pins. This is especially true in race situations (like Fig. 1c, d), where no more pins are possible. For this initial exploration on Plakoto pin databases, we are interested in positions with the following characteristics:

- the side to move has pinned the opponent exactly once inside her bearoff quadrant (points 2–6)[1];
- the opponent has pinned the moving player exactly once;
- no more further pins are possible; in this paper, these no-contact positions are also called *race* positions.

These endgames eventually resolve by one player unpinning his pin, followed by the other player moving his newly freed checker to begin the bearoff. One reason that we are interested in these endgames is that they occur frequently in practice. In an initial 100,000-game self-play experiment with PALAMEDES best neural network Plakoto-5 [7] at the highest settings, we found out that these endgames occur in 14 % of games played.

2.3 Number of Endgame Positions

The number of positions (R) of C checkers residing inside P points can be calculated by the following formula [9]:

$$R = \binom{C+P-1}{C} = \frac{(P+C-1)!}{C!(P-1)!} \tag{1}$$

The number of checkers for the positions of interest is 13 (one checker is pinned by the opponent, and one checker must always be at the pinning point to maintain the pin). Depending on the memory needs of the game-playing program a different number of points (P) can be used. For example for P = 6 (all the non-pinned checkers are under the 6-point, i.e., inside the bearoff quadrant) the total number of positions is 8568 per pin placement. Such a position is shown in Fig. 1c. For the remainder of this paper the assumption will be that all unpinned checkers of the player to move is under the 12 point (P = 12, R = 2496144). A sample position can be seen in Fig. 1d. Note that in both Fig. 1c and d, the position is valid for database retrieval for either player to move.

We have constructed a different database for each possible pin point of the moving player (2–6), so we have 5 databases and 12,480,720 positions in total for the 12-point version. This database is one-sided and corresponds to half the board. If we assume that the opponent has a similar position to the other half, the total possible 2-sided "true" positions that these databases can apply is $12,480,720^2 = 155,768,371,718,400$. If we further assume that the opponent is pinning at the full half of his[2] board (points 13–23) then the total applicable positions are $12,480,720 \times 2,496,144 \times 11 = 342,690,417,780,480$. This number is the lower bound because the endgame character-istics set in the previous section can be met in positions where the opponent player has checkers below the 12 point.

---

[1] No-contact positions where a player has pinned the 1-point (also known as "mana" point) are proven double wins for the pinning player except for the rare cases when the opponent has also pinned the 1-point (tie).
[2] For brevity, 'he' and 'his' whenever 'he or she' and 'his or her' are meant.

3 Algorithm

The goal of a player in the endgame positions that we are interested in is to maintain his pin as long as possible. In essence, the player is playing a mini-game where he tries to maximize the number of moves keeping the pin. Since the game has a chance layer, this goal becomes the maximization of the average distance to unpin. Due to the fact that there is no contact, this metric can be computed using a one-sided database.

3.1 Plakoto Endgame Pin Database Algorithm

The procedure we use is inspired by retrograde analysis [12] where the algorithm starts from a terminal position and works backwards. In our case we do not have terminal positions, but we are starting at a position where all checkers have been moved the furthest. This is the position where all 13 checkers are placed at the last point (point 1). The procedure then works backwards as usual.

Algorithm1. Plakoto endgame pin database creation

---

pinDatabase(p, *pinPlacement*)
  position ← createStartPosition(*pinPlacement*)
  endPos ← createEndPosition(*pinPlacement, p*)
  while *position* is not *endPos* **do**
    saveInDB(hash(*position*), findDistance(*position*))
    increment(*position*)
  end while

  function findDistance(*position*)
    avgDistance ← 0
    for every roll *d* of the 21 possible rolls **do**
      afterStates ← findMoves(*position, d*)
      distances ← readDistancesFromDB(*afterStates*)
      distance ← max(*distances*)
      if *d* is double roll
        avgDistance += *distance*
      else
        avgDistance += 2 * *distance*
      end if
    end for
    return *avgDistance* / 36

The database creation algorithm is shown in Algorithm 1. For every position encountered and all 21 rolls, we find all the legal afterstates, retrieve the distances and return the max distance. The distance of the current position is then calculated as the weighted average of all rolls and stored in the database. The algorithm increments the position and begins the next iteration until all positions are exhausted. The position is incremented in such a way that the resulting afterstates will always have a distance in the database. The only exception is when the roll has no moves, but we can find the distance of this case with a simple recursive operation.

During actual play the database is activated when the position before the roll has the characteristics described in Sect. 2.2. We retrieve the distances of all the afterstates and we select the move which results in the largest distance.

3.2 Storage and Hashing

Important properties for many endgame databases are the storage and the compression mechanisms used. We use a modified version of the hashing function used in [1] to encode the board position to a 32-bit integer. This function is fast, gives a perfect hash, and can be easily decoded for the reversed procedure (int to position). Since the number of records is relatively small we have not made any attempts to compress the database. For the same reason we store the distance value as a double for maximum precision, although it may not be needed. The minimum amount of precision that is acceptable for best play is left for future work. The final database size is 19 MB for the 12-point, and 67 Kb for the 6-point version per pin placement.

4 Discussion

In this section we discuss two interesting challenges with the one-sided databases (Sect. 4.1) and conduct two experiments to evaluate our existing AI in positions from the database (Sect. 4.2).

4.1 Errors in Actual Play with One-Sided Databases

One problem with one-sided databases is that it may give errors in actual play when we take into account the opponent. This is already documented for the one-sided bear-off databases used in backgammon [9]. We identified a class of questionable positions in our pin endgame databases in the very rare situation where the player to move has a high average distance to unpin for all available moves and the opponent is almost ready to unpin. In these cases, because the unpinning of the opponent is almost certain, it may be best for the moving player to prepare for a better placement in the bearoff quadrant instead of continuing to maximize his distance to unpin. However, rollout experiments in 5 samples of such cases have not given evidence that one strategy is better than the other. We believe the problem exists in the one-sided bearoff databases because the bearoff positions are near the end of the game while the questionable pin endgame positions are much further away from terminal. This allows the luck factor to dramatically reduce the effect of the (already small) "error".

4.2 Using the Databases to Evaluate the Neural Networks

Second interesting use of endgame databases (or databases of solved games) is to evaluate existing AI implementations. We conducted experiments with PALAMEDES in two steps using the best neural network (NN) available for Plakoto (1) we checked if the best move of the NN coincided with the best as seen in the databases for all database positions and all possible rolls, and (2) we played 100,000 self-play games with the NN and when a database position was encountered we again compared the move chosen by the NN to the database's optimal move (Table 1). For the first experiment we constructed the opponent position as a mirror of the player to move.

Table 1. Evaluation of PALAMEDES AI in Plakoto pin endgames

Comparison method	Correct moves by the NN (%)
All positions	15 %
Self-play positions	64 %

As can be seen the NN is not selecting the best move 85 % of the time in the first test, however it is doing noticeably better at positions found in practical play. We believe this is normal behavior for the NN to score so low in the first test. The reason is that the self-play procedure used to train the network certainly could not generalize well to all possible cases, most of which are corner cases rarely to be seen in actual play. The result of the second test shows the importance of such databases to enhance the move selection mechanism of the existing AI.

5   Conclusion and Future Work

We have presented an algorithm that created several one-sided endgame databases for the game of Plakoto. The databases are small but can be applied to a huge number of endgame positions. To the best of our knowledge this is the first time that endgame databases are created for the game of Plakoto. We have also shown that the usage of these databases greatly enhances our AI's move selection.

There are several avenues to build upon these results. We mention five of them. The first one is to construct more databases with the same method. We have only built databases for 2–6 pinned points, pinned points 7–18 can be easily created. The second one is building databases with more than one pin per side. The conversion of our algorithm to race endgames where the opponent has pinned more than one checker is straightforward. The third avenue is a more difficult case, viz. is when the moving player has two or more pins.

With the presence of these databases our neural network evaluation function does not need to generalize in these types of positions. We could improve the representation power of our network by retraining the NN without taking into account these endgames.

In the fourth avenue we would also like to explore compression techniques for storage. This will be essential for the creation of larger pin endgame databases.

Acknowledgements.   The authors would like to thank the anonymous referees for their useful comments and suggestions that contributed to improving the final version of the paper.

References

1. Benjamin, A., Ross, A.M.: Enumerating backgammon positions: the perfect hash. Interface: Undergraduate Res. Harvey Mudd Coll. **16**(1), 3–10 (1996)
2. Bowling, M., Burch, N., Johanson, M., Tammelin, O.: Heads-up limit hold'em poker is solved. Science **347**(6218), 145–149 (2015)
3. Ciancarini, P., Favini, G.P.: Solving Kriegspiel endings with brute force: the case of KR vs. K. In: van den Herik, H., Spronck, P. (eds.) ACG 2009. LNCS, vol. 6048, pp. 136–145. Springer, Heidelberg (2010)
4. Fang, H.-R., Hsu, T.-S., Hsu, S.-C.: Construction of Chinese chess endgame databases by retrograde analysis. In: Marsland, T., Frank, I. (eds.) CG 2001. LNCS, vol. 2063, pp. 96–114. Springer, Heidelberg (2002)
5. Gasser, R.: Solving nine men's morris. Comput. Intell. **12**(1), 24–41 (1996)
6. Nalimov, E.V., Haworth, G.M., Heinz, E.A.: Space-efficient indexing of chess endgame tables. ICGA J. **23**(3), 148–162 (2000)
7. Papahristou, N., Refanidis, I.: On the design and training of bots to play backgammon variants. In: Iliadis, L., Maglogiannis, I., Papadopoulos, H. (eds.) Artificial Intelligence Applications and Innovations. IFIP AICT, vol. 381, pp. 78–87. Springer, Heidelberg (2012)
8. Romein, J.W., Bal, H.E.: Solving awari with parallel retrograde analysis. Computer **36**(10), 26–33 (2003)
9. Ross, A.M., Benjamin, A.T., Munson, M.: Estimating winning probabilities in backgammon races. In: Ethier, S.N., Eadington, W.R. (eds.) Optimal Play: Mathematical Studies of Games and Gambling, pp. 269–291. Institute for the Study of Gambling and Commercial Gaming, University of Nevada, Reno (2007)
10. Schaeffer, J., Bjornsson, Y., Burch, N., Lake, R., Lu, P., Sutphen, S.: Building the checkers 10-piece endgame databases. In: van den Herik, H.J., Iida, H., Heinz, E.A. (eds.) Advances in Computer Games: Many Games, Many Challenges, pp. 193–210. Kluwer Academic Publishers, Dordrecht (2003)
11. Schaeffer, J., Burch, N., Björnsson, Y., Kishimoto, A., Müller, M., Lake, R., Lu, P., Sutphen, S.: Checkers is solved. Science **317**, 1518–1522 (2007)
12. Thompson, K.: Retrograde analysis of certain endgames. ICCA J. **9**(3), 131–139 (1986)

Reducing the Seesaw Effect with Deep Proof-Number Search

Taichi Ishitobi[1]([✉]), Aske Plaat[2], Hiroyuki Iida[1], and Jaap van den Herik[2]

[1] Japan Advanced Institute of Science and Technology, Nomi, Japan
{itaichi,iida}@jaist.ac.jp
[2] Leiden Institute of Advanced Computer Science, Leiden, The Netherlands
{aske.plaat,jaapvandenherik}@gmail.com

Abstract. In this paper, DeepPN is introduced. It is a modified version of PN-search. It introduces a procedure to solve the seesaw effect. DeepPN employs two important values associated with each node, viz. the usual proof number and a *deep value*. The deep value of a node is defined as the depth to which each child node has been searched. So, the deep value of a node shows the progress of the search in the depth direction. By mixing the proof numbers and the deep value, DeepPN works with two characteristics, viz., the best-first manner of search (equal to the original proof-number search) and the depth-first manner. By adjusting a parameter (called R in this paper) we can choose between best-first or depth-first behavior. In our experiments, we tried to find a balance between both manners of searching. As it turned out, best results were obtained at an R value in between the two extremes of best-first search (original proof number search) and depth-first search. Our experiments showed better results for DeepPN compared to the original PN-search: a point in between best-first and depth-first performed best. For random Othello and Hex positions, DeepPN works almost twice as good as PN-search. From the results, we may conclude that Deep Proof-Number Search outperforms PN-search considerably in Othello and Hex.

1 Introduction

Proof Number Search (PN-search) was developed by Allis et al. in 1994 [16]. It is one of the most powerful algorithms for solving games and complex endgame positions. PN-search focuses on an AND/OR tree and tries to establish the game-theoretical value in an efficient way, in a greedy least-work-first manner. Each node has a proof number (pn) and disproof number (dn). This idea was inspired by McAllister's concept of conspiracy numbers, the number of leaves that need to change their value for a node to change its value [10]. A proof number shows the scale of difficulty in proving a node, a disproof number does analogously for disproving a node. PN-search tries to expand a most-proving node, which is the most efficient one for proving (disproving) a node. PN-search is a best-first search in the sense that it follows a least-work-first order.

The success of PN-search prompted researchers to create many derivatives of PN-search, e.g., PN* [17], PDS [6] and df-pn [7]. The history of these algorithms

© Springer International Publishing Switzerland 2015
A. Plaat et al. (Eds.): ACG 2015, LNCS 9525, pp. 185–197, 2015.
DOI: 10.1007/978-3-319-27992-3_17

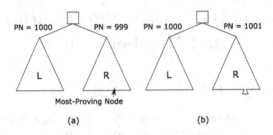

Fig. 1. An example of the Seesaw effect: (a) An example game tree (b) Expanding the most-proving node

is described by Kishimoto et al. [2]. Using one of these algorithms, many games and puzzles were solved, e.g., Checkers [13,14], and Tsume-shogi (the mating problem of Japanese chess) [17]. The algorithms used many well-known techniques such as transposition tables. On the one hand researchers sought to benefit from large computer power and memory [15], and on the other hand, researchers worked at some of the problems of PN-search, such as the seesaw effect (see Sect. 2).

In this paper, we propose a new proof-number algorithm called Deep Proof-Number Search (DeepPN). DeepPN tries to solve the seesaw problem with a different approach together with iterative deepening. In the experimental section, we use endgame positions of Othello and the game of Hex as benchmarks, and try to measure the performance of DeepPN.

The remainder of this paper is as follows. We briefly summarize the details of the seesaw effect in Sect. 2. The algorithm of DeepPN and its performance are discussed in Sect. 3. In Sect. 4 we give our conclusion and suggestions for future work.

2   The Seesaw Effect

The derivatives of PN-search address many issues of the algorithm, and have been used to solve many games. However, there are still some problems with PN-search that remain. Pawlewicz and Lew [12], and Kishimoto et al. [3,4] showed one such weak point, namely that of df-pn. The weak point has been named the Seesaw-effect by Hashimoto [11]. Below we provide a short sketch.

To explain the seesaw effect, we show an example in Fig. 1. In Fig. 1a, a root node has two large subtrees. The size of both subtrees is almost the same. Further assume that the proof number of subtree L is larger than the proof number of subtree R. In this case, PN-search will focus on subtree R, will continue the searching, and will expand the most-proving node. When PN-search expands a most-proving node, the shape of a game-tree changes as shown in Fig. 1b. By expanding the most-proving node, the proof number of subtree R becomes larger than the proof number of subtree L. Because of this, the position of the most-proving tree changes from subtree R to subtree L. Similarly, when the search expands the most-proving node in subtree L, the proof number of subtree L changes to a larger value than the proof number of subtree R. Thus, the search switches its focus from subtree L to subtree R. This changing continues to go back and forth and looks like a seesaw. Therefore, it is named the Seesaw effect. The Seesaw effect happens when the two trees are almost equal in size.

If the Seesaw effect occurs, the performance of PN-search and df-pn deteriorates significantly [7]. Df-pn usually tries to search efficiently by staying around a most-proving node as in Fig. 1a. However, when the seesaw effect occurs, df-pn should go back to the root node, and switch focus to another subtree and start to find a new most-proving node existing in subtree L. If the seesaw effect occurs frequently, the performance of df-pn becomes close to that of PN-search, because df-pn loses the power of its depth-first behavior.

For PN-search, the algorithm uses the proof numbers to search efficiently in a best-first manner. If the seesaw effect occurs frequently, PN-search will concentrate alternatively on one subtree. PN-search will then expand subtrees L and R equally and it cannot reach the required depth. In games which need to reach a large fixed depth for solving, this effect works strongly against efficiency.

The causes of the seesaw effect are mostly (1) the shape of the game tree and (2) the way of searching. Concerning the shape of game tree, there are two characteristics: (1a) a tendency for the number of child nodes to become equal and (1b) many nodes with equal values exist deep down in a game tree. In (1a), if the number of a child node in each node becomes almost the same, then the seesaw effect may occur easily. For (1b), this is the case in games such as Othello and Hex. In many cases, these games need to search a large fixed number of moves before settling, and it is difficult to assess upon a win, loss, or draw before a certain number of moves has been played. In the game tree of these games, the nodes can establish their value after a certain depth has been searched. Thus, when the seesaw effect occurs and the search cannot reach the required depth, it cannot determine the status of the subtrees. Instead of seesawing between subtrees, the search should stick with one subtree and search more deeply. A game tree that has these characteristics is called a *suitable tree* by Hashimoto. Games such as Othello, Hex and Go are able to build up a suitable tree easily. For (2), the way of searching, i.e., the best-first manner causes the seesaw effect. The most-proving node of PN-search and df-pn is determined using proof numbers. Thus, in the Fig. 1, df-pn has to go back to the root node time and again, and PN-search and df-pn cannot reach a required depth in the subtree.

One solution for the seesaw effect is the "1 + ϵ trick" proposed by Pawlewicz and Lew [12]. They focused on df-pn and changed the term for calculating the threshold. To paraphrase their explanation, they add a margin determined by ϵ to the thresholds. This margin is calculated by the size of other subtrees, and it is recalculated in each seesaw. By the added margin for the thresholds, df-pn can reach nodes in a specific branch more deeply than the original algorithm. Hence, the frequency of the seesaw effect is reduced. Consequently, df-pn with the 1 + ϵ trick works better than the original df-pn. However, we believe that this trick has at least three problems. First, the trick breaks a rule about the most-proving node. The original thresholds keep the definition of most-proving node, but $1 + \epsilon$ just adds a margin to the thresholds. Second, if the game tree changes become too large, then also the margin becomes too large, because the margin is calculated by the size of the other subtree. On the one hand, a large margin can reduce the frequency of the seesaw effects, on the other hand, if a subtree to be searched is found not to lead to any result, then the search cannot change the subtree until reaching that margin. Third, the $1 + \epsilon$ trick only reduces the frequency of the seesaw effect and does not completely solve the problem.

3   DeepPN

In this section, we explain a new algorithm based on proof numbers named Deep Proof-Number Search (DeepPN). DeepPN is modeled after the original PN-search, and all nodes have proof numbers and dis-proof numbers. Additionally, for DeepPN, each node is assigned also a so-called *deep value*. The deep values are determined and updated by the terminal node analogously to the proof and dis-proof numbers. DeepPN has been designed: (1) to combine best-first and depth-first search, and (2) to try and solve the problem of the Seesaw effect. For evaluating the performance of DeepPN, we use endgame positions of Othello and Hex.

3.1   The Basic Idea of DeepPN

In the original PN-search, the most-proving node is defined as follows [16].

Definition. For any AND/OR tree T, a most-proving node of T is a frontier node of T, which by obtaining the value true reduces T's proof number by 1, while by obtaining the value false reduces T's disproof number by 1.

This definition implies that the most-proving node sometimes exists in a plural form in a tree, i.e., there are many fully equivalent most-proving nodes. For example, if the child nodes have the same proof or disproof number then both subtrees have each a most-proving node. The situation that the child node has the same proof (disproof) number in an OR (AND) node is called a tie-break situation. Now, we have the question about which most-proving node is the best for calculating the game-theoretical value. PN-search chooses the leftmost node with the smallest proof (disproof) number, also in a tie-break situation. In particular, the proof and disproof number do not take other information into account, and therefore PN-search cannot choose a more favorable most-proving node in a tie-break situation.

Determining the best most-proving node in a tie-break situation is a difficult task, because the answer depends on many aspects of the game. However, when focusing on games which build up a suitable tree, we may develop some solutions. In a suitable tree, the "best" most-proving node is indicated by its depth number. Let us look at the example (given in Fig. 2).

This game tree is based on Othello. The game end is shown by "Game End" in Fig. 2. All level-two nodes are most-proving nodes, because the proof numbers of child nodes under the root node are the same (i.e., 2). So, we have a tie-break

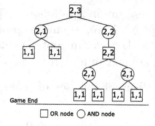

Fig. 2. An example of a suitable tree for an Othello end-game position. This game tree has a uniform depth of 4, and the terminal nodes are reached at game end.

situation. Now, in the next search step, PN-search will focus on the most-proving node that exists in the left side as produced by the original PN-search algorithm. However, if the search focuses immediately on the most-proving node of the right side, then the search will be more efficient, because the nodes on the left side do not reach the game end and their value cannot be found yet. In contrast, nodes that exist at the right side reach the game end, and if we try to expand these nodes, then the game value of each node is known. In this example, we follow the idea that a most-proving node in the deepest tree of a suitable game tree, is the best.

To test this idea, we performed a small experiment. We prepared an original PN-search and a modified PN-search. In a tie-break situation, PN-search focuses on a most-proving node that exists in the leftmost node, and the modified PN-search focuses on the *deepest* most-proving node. For checking performance, we prepared 100 Othello endgame positions. The performance of the modified PN-search is better than the results of the original PN-search (about 10 % reduction). These results suggest that the *deepest* most-proving node works advantageously for finding the game-theoretical value.

In addition, the example of Fig. 2 shows the essence of the seesaw effect. If the game end exists and has a depth of more than 4, then the search for a proof number goes back and forth between the two subtrees. Even if the game end is of depth 4, then the search that focuses on the right subtree will change its focus on the left subtree. But, when modifying PN-search, the small seesaw effect is suppressed. This phenomenon of modifying PN-search suggests a new heuristic. The search depth of nodes can be used for solving the seesaw effect in a suitable game tree. In fact, this is what the $1 + \epsilon$ trick [12] in effect tries to accomplish, to stay deep in a suitable game tree. Now, let us try to think of a new technique. For instance, consider the moves that the modified PN-search plays when finding the deepest most proving node. We noticed that these moves combined best-first with depth-first behavior. The modified PN-search works in a best-first manner, and in a tie-break situation, PN-search works depth-first for the most-proving nodes. Depending on how often tie-breaks occur, the algorithm works more frequently best-first than depth-first. The resulting improvement, when measured in number of iterations and nodes leads to a small result. Thus, we will design a new algorithm that can change the ratio of best-first manner and depth-first manner. Its description is as follows. This system is named Deep Proof-Number Search (DeepPN). Here, $n.\phi$ means proof number in an OR node and disproof number in an AND node. In contrast, $n.\delta$ means proof number in an AND node and disproof number in an OR node.

1. The proof number and disproof number of node n are now calculated as follows.

$$n.\phi = \begin{cases} n.pn \text{ (n is an OR node)} \\ n.dn \text{ (n is an AND node)} \end{cases}$$

$$n.\delta = \begin{cases} n.dn \text{ (n is an OR node)} \\ n.pn \text{ (n is an AND node)} \end{cases}$$

2. When n is a terminal node
   (a) When n is proved (disproved) and n is an OR (AND) node, i.e., OR wins

$$n.\phi = 0, \ n.\delta = \infty$$

   (b) When n is disproved (proved) and n is an AND (OR) node, i.e., OR does not win

$$n.\phi = \infty, \ n.\delta = 0$$

   (c) When n is unsolved, i.e., its value is unknown

$$n.\phi = 1, \ n.\delta = 1$$

   (d) When n is terminal node, then n has deep value

$$n.deep = \frac{1}{n.depth} \tag{1}$$

3. When n is an internal node
   (a) The proof and disproof number are defined as follows

$$n.\phi = \min_{n_c \in \text{ children of } n} n_c.\delta$$

$$n.\delta = \sum_{n_c \in \text{ children of } n} n_c.\phi$$

   (b) The deep values, DPN(n) and n.deep are defined as follows.

$$n.deep = n_c.deep \text{ where } n_c = \operatorname*{arg\,min}_{n_i \in \text{ unsolved children}} DPN(n_i) \tag{2}$$

$$DPN(n) = (1 - \frac{1}{n.\delta})R + n.deep(1 - R) \quad (0.0 \le R \le 1.0) \tag{3}$$

The proof and disproof number are the same as in the original PN-search. The improvement is the new term, i.e., the concept of the *deep* value. The deep value in a terminal node is calculated by formula (1). The deep value is designed to decrease inversely with depth. In an internal node, calculating the deep value has only a limited complexity. First, we define a function named DPN (see formula 3). DPN has two features: (a) $n.\delta$ is normalized and designed to become larger according to the growth of $n.\delta$ and (b) a fixed parameter R is chosen. R has a value between 0.0 and 1.0. If R is 1.0 then DeepPN works the same as PN-search, and if R is 0.0 then DeepPN works the same as a primitive depth-first search. Therefore, the normalized δ fulfills the role of best-first guide and the *deep* acts as a depth-first guide. This means that by changing the value of R, the ratio of best-first and depth-first search of DeepPN can be adjusted. Second, in an internal node, the deep value is updated by its child nodes using formula (2). The deep value of node n is decided by a child node n_c which has smallest $DPN(n_c)$. A point to notice is that the updating value is only *deep*, not $DPN(n_c)$. Additionally, when n_c is solved, then the deep value of n_c is ignored in arg min.

In DeepPN, an expanding node in each iteration is chosen as follows.

$$select\_expanding\_node(n) := \underset{n_c \in \text{ children of } n \text{ except solved}}{\arg\min} DPN(n_c) \qquad (4)$$

This sequence is repeated until the terminal node is reached. That terminal node is the node that is to be expanded. If $R = 1.0$, then this expanding node is the most-proving node.

3.2 Performance with Othello

For measuring the performance of DeepPN, we prepared a solver using the DeepPN algorithm and Othello endgame positions. We configured a primitive DeepPN algorithm for investigating the effect of DeepPN only, without any supportive mechanisms such as transposition tables and ϵ-thresholds. We prepared 1000 Othello endgame positions. They are constructed as follows. The positions are taken from the 8×8 board. We play 44 legal moves at random from the begin position. This implies that 48 squares from the 64 are covered. So, the depth of the full tree to the end is 16.

In all our experiments DeepPN is applied to these 1000 endgame positions. Our focus is the behavior of R (see formula (3)). For $R = 1.0$, DeepPN works the same as PN-search and shows the same results. For $R = 0.0$, DeepPN works the same as a primitive depth-first search. When R is between 1.0 to 0.0, then DeepPN behaves as a mix between best-first and depth-first. We changed R from 1.0 to 0.0 by decrements of 0.05. We focus on the values of two concepts, viz. the number of iterations and the number of nodes. The number of iterations is given by counting the number of traces of finding the most-proving node from the root node. This value indicates an approximate execution time unaffected by the specifications of a computer. The number of nodes is an indication of the total number of nodes that are expanded by the search. This value is an approximation of the size of memory needed for solving. We show the results in Fig. 3.

Figure 3a shows the variation of (1) the number of iterations and (2) the number of nodes. Each point is the mean value calculated from the results of

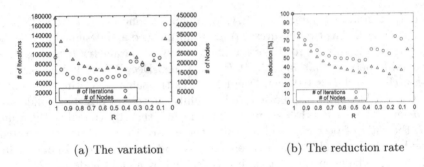

(a) The variation   (b) The reduction rate

Fig. 3. Othello: The number of Iterations and Nodes, for the variation (left) and the reduction rate (right) $R = 1.0$ is PN-search, $R = 0.0$ is depth-first search, and $1.0 > R > 0.0$ is DeepPN. Lower is better.

1000 Othello endgame positions. $R = 1.0$ shows the results of PN-search, and this value is the base for comparison. As R goes to 0.8, the number of iterations and nodes decrease almost by half. From $R = 0.8$ to 0.6, the number of iterations stops decreasing, but the number of nodes decreases slowly. From $R = 0.6$ to 0.4, the decrease stops, and the number of iterations starts increasing again slowly. In $R = 0.35$, both numbers increase rapidly. We see that for R of around 0.4, the balance between depth-first and best-first behavior appears to be optimal. We surmise that DeepPN is stuck in one subtree and cannot get away since the algorithm is too strongly depth-first. For $R = 0.35$ to 0.2, the number of iterations and nodes is decreasing. Around $R = 0.2$, the balance was broken again, and is decreasing towards 0.1. Finally, DeepPN performs worse when R approaches 0.0 closely. In $R = 0.0$, almost no Othello end game position can be solved, and this value is omitted from Fig. 3a.

In Fig. 3a, the scale of the number of iterations and nodes are different. To ease our understanding, Fig. 3b shows the amount of the reduction rate. This reduction rate is normalized by the result of PN-search, i.e., the reduction rate of $R = 1.0$ is 100 %. Each point is the mean value of the reduction rate calculated by the results of 1000 Othello endgame positions. The results of Fig. 3b show almost the same characteristics as Fig. 3a. There is a different point where the number of iterations decreases after $R = 0.8$ and the number of nodes decreases after $R = 0.6$. In Fig. 3b, the number of iterations decreased about 50 % in $R = 0.4$ and the number of nodes decreased about 35 % in $R = 0.4$. Thus, DeepPN reduced the number of iterations ($\approx$ time) to half and the number of nodes ($\approx$ space) to one-third. In $R = 0.05$, the number of iterations increased to over 100 %, which is not shown.

Finally, we show two graphs about the changes in *reducing* and *increasing* cases in Othello endgame positions in Fig. 4. Please note that in Fig. 4 we showed the number of iterations and number of nodes.

The plots for reducing cases give the number of Othello endgame positions which are solved efficiently compared to PN-search, i.e., the reduction rate is under 100 %. In contrast, the plots for increasing cases give the number of

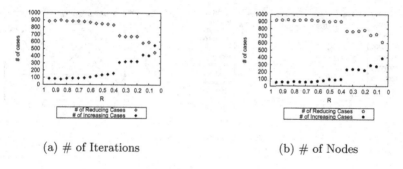

(a) # of Iterations   (b) # of Nodes

Fig. 4. Othello: The changes of Reducing and Increasing Cases for # of Iterations and # of Nodes

Othello endgame positions that have a reduction rate over 100 %. The vertical axis shows the number of Othello endgame positions. Figure 4a shows the number of iterations by which the reducing cases decrease slowly from $R = 0.95$ to 0.4. Likewise, for number of nodes the graph decreases slowly from $R = 0.95$ to 0.4. Around $R = 0.4$, the trend is broken, and the number of increasing cases increases rapidly. From $R = 0.35$ to 0.2 and from 0.15 to 0.1, the number of cases does not change much. This result indicates that the reason of decreasing from $R = 0.35$ to 0.2 is shown in Fig. 3a and b. As the number of cases is not changed, the decreasing number of iterations and nodes of the Othello endgame positions are caused by reducing cases. In brief, some Othello end positions can be handled efficiently as R is reduced. But, for some Othello endgame positions a changing R causes an increase. Therefore, Othello endgame positions can be categorized in relation to R. The first group belongs to $R = 0.95$ to 0.05. This group does not react to changes in R, they do not switch between the reducing case and increasing case. We can see this group clearly from $R = 0.95$ to 0.40. The second group belongs to $R = 0.35$ to 0.2. This group fitted from $R = 0.95$ to 0.4, and they could not keep efficiency work after $R = 0.4$. The third group belongs to $R = 0.15$ to 0.1, and the characteristics of this group are the same as for the second group. In either group, the cases are not efficiently close to $R = 0.0$.

The question remains when DeepPN works most efficiently in the Othello endgame position for 16-ply. The answer depends on the group of Othello endgame positions. However, if we have to choose the best R, then a value of around 0.65 is a good compromise for most cases.

3.3 Performance with Hex

For measuring the performance of DeepPN, we also prepared a solver for Hex. Similarly to the experiments of Othello, we created a primitive DeepPN algorithm for checking the effect of DeepPN only. The Hex program is a simple program that does not have any other mechanisms such as an evaluation function. Our Hex program uses a 4×4 board (called Hex(4)), and tries to solve that board using DeepPN. Our focus is on the behavior of R (see formula 3).

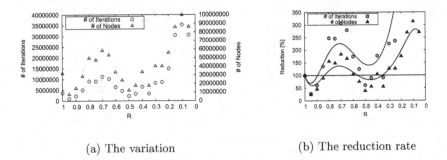

(a) The variation   (b) The reduction rate

Fig. 5. Hex: # of Iterations and # of Nodes for Hex(4). $R = 1.0$ is PN-search, $R = 0.0$ is depth-first search, and $1.0 > R > 0.0$ is DeepPN. Lower is better.

Concerning the characteristics of R, we refer to Sects. 3.1 and 3.2. We changed R from 0.0 to 1.0 by 0.05, and tried to solve Hex(4) 10 times in each R. The legal moves of Hex are sorted randomly in every configuration, viz. there is the possibility that each result is different. The results in each R are calculated by the average of the 10 experiments. Next we focused on two concepts: (1) number of iterations and (2) number of nodes. About the characteristics of both values, we refer to Sect. 3.2. The experimental data are given in Fig. 5.

Figure 5a shows the changes in the number of iterations and nodes. We can see that the results of DeepPN decrease (improve) in some positions compared by PN-search. This is not the case for $R = 0.0$, because we cannot solve Hex(4) for this R if we limit ourselves to 500 million nodes. For ease of understanding, we prepare another graph in Fig. 5b. There we show the reduction rates normalized by the result of PN-search, i.e., the result of PN-search has 100 % reduction rate.

Figure 5b shows that the number of iterations and nodes is reduced by a 30 % reduction rate between $R = 0.95$ and $R = 0.5$. The result has two downward curves: from $R = 1.0$ to 0.7 and from $R = 0.7$ to $R = 0.0$. The first curve starts from $R = 1.0$ and decreases toward 0.95. After $R = 0.95$, the results start to increase and grow to over 100 % after 0.85. The second curve starts from $R = 0.7$ and the results starts to decrease again. At around $R = 0.5$, the results reach about 50 %. Finally, the results are increasing again toward $R = 0.0$, just as for Othello.

For understanding the details of how DeepPN works around $R = 0.95$, we tried to change R by 0.1 between from 1.0 to 0.9. The results are shown in Fig. 6.

By looking at the results, we can see that DeepPN works almost twice as good as PN-search from $R = 0.99$ to 0.95. From $R = 0.95$ to 0.90, we have a small curve like Fig. 5b.

In Hex(4), the optimum value of R is around $R = 0.95$ (and perhaps $R = 0.5$). We can see that depth-first does not work so well for Hex(4) as it does for Othello, although there is an improvement over pure best-first.

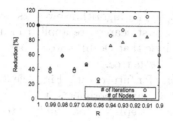

Fig. 6. Hex: The detail of Fig. 5b. This figure is zoomed in for $1.0 \leq R \leq 0.9$. The lower is better.

3.4 Discussion

DeepPN works efficiently in 16-ply Othello endgame positions, and in Hex(4). It can reduce the number of iterations and nodes almost by half compared to PN-search. It must be noted that the optimum balance of R is different in each game and for each size of game tree. We can see that for both games a certain amount of depth-first behavior is beneficial, but the changes are not the same. The precise relation is a topic of future work.

Both in Othello endgame positions and in Hex(4), we encountered positions that showed increasing (worse) results. We suspect that a reason for this problem may be (1) the holding problem and (2) the length of the shortest correct path. Concerning (1), the depth-first search can remain stuck in one subtree (holding on to the subtree). If this holding subtree cannot find the game-theoretical value, then the number of iterations and nodes become meaningless. When DeepPN employed a strong depth-first manner, then we found many increasing (worse) results in Othello endgame positions. Also, in Hex(4), DeepPN cannot work efficiently around $R = 0.0$. Finding an optimal R is a topic of future work.

Concerning (2), the problem is related to (1). In Othello, the shortest correct path is almost the same for each position, because Othello has a fixed number of depth to the end. However, in Hex(4), the shortest winning path may exist before a depth of 16. If we happen to find a balance between depth and best-first, then DeepPN will change the subtree on which it focuses on time. For example, when $R = 0.95$, then DeepPN quickly finds the shortest path. But after $R = 0.95$, DeepPN misses that path and arrives in regions that are more deeply in the trees. Finding a good value of R in Hex is more difficult than in Othello.

4   Conclusion and Future Works

In this work, we proposed a new search algorithm based on proof numbers, named DeepPN. DeepPN has three values (pn, dn, *deep*) and a single parameter, R, that allows a choice between depth-first and best-first behavior. DeepPN employs two types of values, viz. proof numbers and deep values which register the depth of nodes. For measuring the performance of DeepPN, we tested DeepPN on solving Othello endgame positions and on the game of Hex. We achieved two indicative results in Othello and Hex. The algorithm owes its success to formula (3) in which best-first and depth-first search are applied in a "balanced" way. From there the results may conclude that DeepPN works better than PN-search in the games which build up a *suitable* tree.

We have four main topics for future work. First, we have to investigate how to find a good balance for R. In our experiments, the best results are produced by different values of R. Second, DeepPN is too primitive to solve complex problems. Df-pn+ and the use of a transposition table give a good hint for this problem. The idea of deep value may also be applied to other ways of searching. Third, we need to find a reason about the deterioration of search results in Othello endgame positions and in the game of Hex. By investigating this problem, we expect that DeepPN can become a quite useful algorithm for explaining the intricacies.

The fourth topic is further away. Previous work on depth-first and best-first minimax algorithms used null-windows around the minimax value to guide the search in a best-first manner [8,9]. There, a relation between the SSS* and the Alpha-beta algorithm was found. In this work, we focus on best-first behavior guided by the *size* of the search tree, as is the essence of proof-number search. It will be interesting to see if both kinds of best-first behavior can be combined in future work in a new kind of conspiracy number search.

Acknowledgements. This research is funded by a grant from the Japan Society for the Promotion of Science, in the framework of the Grant-in-Aid for Challenging Exploratory Research (grant number 26540189).

References

1. Kishimoto, A., Müller, M.: About the completeness of depth-first proof-number search. In: van den Herik, H.J., Xu, X., Ma, Z., Winands, M.H.M. (eds.) CG 2008. LNCS, vol. 5131, pp. 146–156. Springer, Heidelberg (2008)
2. Kishimoto, A., Winands, M., Müller, M., Saito, J.-T.: Game-tree search using proof numbers: the first twenty years. ICGA J. **35**(3), 131–156 (2012)
3. Kishimoto, A., Muller, M.: Search versus knowledge for solving life and death problems in go. In: Twentieth National Conference on Artificial Intelligence (AAAI 2005), pp. 1374–1379 (2005)
4. Kishimoto, A.: Correct and Efficient Search Algorithms in the Presence of Repetitions, Ph.D. thesis, University of Alberta (2005)
5. Nagai, A.: Df-pn Algorithm for Searching AND/OR Trees and Its Applications. Ph.D. thesis, Dept. of Information Science, University of Tokyo, Tokyo (2002)
6. Nagai, A.: A new AND/OR tree search algorithm using proof number and disproof number. In: Proceedings of Complex Games Lab Workshop, pp. 40–45. ETL, Tsukuba (1998)
7. Nagai, A.: A new depth-first search algorithm for AND/OR trees. M.Sc. thesis, Department of Information Science, The University of Tokyo, Japan (1999)
8. Plaat, A., Schaeffer, J., Pijls, W., de Bruin, A.: Best-first and depth-first minimax search in practice. In: Proceedings of Computer Science in the Netherlands, pp. 182–193 (1995)

9. Plaat, A., Schaeffer, J., Pijls, W., de Bruin, A.: SSS* = Alpha-beta + TT. Technical report 94–17, University of Alberta, Edmonton, Canada (1994)
10. McAllester, D.: Conspiracy numbers for min-max search. Artif. Intell. **35**(1), 287–310 (1988)
11. Hashimoto, J.: A Study on Game-Independent Heuristics in Game-Tree Search. Ph.D. thesis, School of Information Science, Japan Advanced Institute of Science and Technology (2011)
12. Pawlewicz, J., Lew, L.: Improving depth-first PN-search: 1 + ϵ trick. In: van den Herik, H.J., Ciancarini, P., Donkers, H.H.L.M.J. (eds.) CG 2006. LNCS, vol. 4630, pp. 160–171. Springer, Heidelberg (2007)
13. Schaeffer, J., Björnsson, Y., Burch, N., Kishimoto, A., Müller, M., Lake, R., Lu, P., Sutphen, S.: Checkers is solved. Science **317**(5844), 1518–1522 (2007)
14. Schaeffer, J.: Game over: black to play and draw in checkers. ICGA J. **30**(4), 187–197 (2007)
15. Hoki, K., Kaneko, T., Kishimoto, A., Ito, T.: Parallel dovetailing and its application to depth-first proof-number search. ICGA J. **36**(1), 22–36 (2013)
16. Allis, L.V., van der Meulen, M., van den Herik, H.J.: Proof-number search. Artif. Intell. **66**(1), 91–124 (1994)
17. Seo, M., Iida, H., Uiterwijk, J.W.H.M.: The PN*-search algorithm: application to tsume-shogi. Artif. Intell. **129**(4), 253–277 (2001)
18. Winands, M.H.M.: Informed Search in Complex Games., Ph.D. thesis, Maastricht University, The Netherlands (2004)
19. Ueda, T., Hashimoto, T., Hashimoto, J., Iida, H.: Weak proof-number search. In: van den Herik, H.J., Xu, X., Ma, Z., Winands, M.H.M. (eds.) CG 2008. LNCS, vol. 5131, pp. 157–168. Springer, Heidelberg (2008)

Feature Strength and Parallelization of Sibling Conspiracy Number Search

Jakub Pawlewicz[1] and Ryan B. Hayward[2]($\boxtimes$)

[1] Institute of Informatics, University of Warsaw, Warsaw, Poland
pan@mimuw.edu.pl
[2] Computing Science, University of Alberta, Edmonton, Canada
hayward@ualberta.ca

Abstract. Recently we introduced Sibling Conspiracy Number Search —
an algorithm based not on evaluation of leaf states of the search tree but,
for each node, on relative evaluation scores of all children of that node —
and implemented an SCNS Hex bot. Here we show the strength of SCNS
features: most critical is to initialize leaves via a multi-step process. Also,
we show a simple parallel version of SCNS: it scales well for 2 threads
but less efficiently for 4 or 8 threads.

1 Introduction

Call a heuristic function *local* if it accurately compares the strength of sibling
moves in the search tree. Recently we introduced Sibling Conspiracy Number
Search [19], an algorithm designed for such a heuristic[1], and implemented DEEP-
HEX, an SCNS Hex bot. In this paper we show feature strength of our SCNS
Hex bot, and also describe a parallel implementation.

2 Conspiracy Number Search

In 2-player game search, CNS has shown promise in chess [13–15,17,23,24] and
shogi [11]. CNS can be viewed as a generalization of PNS (Proof Number Search).

PNS is used in two-player zero-sum games. One player is *us*, the other is
them or *opponent*. Value *true* (*false*) is a win for us (them). We (they) move at
an or-node (and-node). PNS is hard to guide with an evaluation function, as
leaves have only two possible values [1,4,12,18,22,25,26]. One can extend PNS
by allowing a leaf to have any rational value, with $+(-)\infty$ for win(loss). If a
leaf is terminal, its value is the actual game value; if not terminal, its value can
be assigned heuristically. We (they) want to maximize (minimize) value. A node
from which we (they) move is a *max-node* (*min-node*). Internal node values
are computed in minimax fashion. MINIMAX(n) denotes the minimax value of
node n.

---

[1] Hex has a good local heuristic. Shannon built an analogue circuit to play the connection game Bridg-it, with moves scored by voltage drop [7]. Adding links between
virtual connected cells [2] improves the heuristic, which is reliable among siblings [9].

© Springer International Publishing Switzerland 2015
A. Plaat et al. (Eds.): ACG 2015, LNCS 9525, pp. 198–209, 2015.
DOI: 10.1007/978-3-319-27992-3_18

PNS is computed using the two final values (true/false) and a temporary value (unknown) assigned to non-terminal leaves. The (dis)proof number measures how difficult it is to change from unknown to true (false). Rather than numbers, we use functions to represent the extended set of values denoted by $V = \{-\infty\} \cup \mathbb{R} \cup \{+\infty\}$.

Definition 1. The function $p_n : V \mapsto \mathbb{N}_0 = \{0, 1, 2, \ldots\}$ is a *proof function* if, for all $v \in V$, $p_n(v)$ is the minimum number of leaves in the subtree rooted at n that must change value so that $\text{MINIMAX}(n) \geq v$. Similarly, $d_n : V \mapsto \mathbb{N}_0$ is a *disproof function* if, for all $v \in V$, $d_n(v)$ is the minimum number of leaves in the subtree rooted at n that must change value so that $\text{MINIMAX}(n) \leq v$.

Rather than storing (dis)proof numbers at each node, we store (dis)proof functions, computed recursively: If n is a leaf and x is its value (heuristic or actual) then

$$p_n(v) = \begin{cases} 0 & \text{if } v \leq x \\ 1 & \text{if } v > x \text{ and } n \text{ is non-terminal} \\ +\infty & \text{if } v > x \text{ and } n \text{ is terminal}, \end{cases}$$

$$d_n(v) = \begin{cases} 0 & \text{if } v \geq x \\ 1 & \text{if } v < x \text{ and } n \text{ is non-terminal} \\ +\infty & \text{if } v < x \text{ and } n \text{ is terminal}, \end{cases}$$

(1)

otherwise, for every $v \in V$,

$$p_n(v) = \min_{s \in \text{children}(n)} p_s(v), \quad d_n(v) = \sum_{s \in \text{children}(n)} d_s(v) \quad \text{if } n \text{ is or-node,}$$

$$p_n(v) = \sum_{s \in \text{children}(n)} p_s(v), \quad d_n(v) = \min_{s \in \text{children}(n)} d_s(v) \quad \text{if } n \text{ is and-node.}$$

(2)

(Dis)Proof functions can be propagated up from leaves; f can be stored as an array of possible $f(v)$ for each v. For each node n

(i) p_n is a non-decreasing staircase function, and
    d_n is a non-increasing staircase function.
(ii) $\text{MINIMAX}(n)$ is the meet point of p_n and d_n, i.e.,:

$$\begin{array}{lll} p_n(v) = 0, & d_n(v) > 0 & \text{for } v < \text{MINIMAX}(n), \\ p_n(v) = 0, & d_n(v) = 0 & \text{for } v = \text{MINIMAX}(n), \\ p_n(v) > 0, & d_n(v) = 0 & \text{for } v > \text{MINIMAX}(n). \end{array}$$

See Fig. 1. Following McAllester, the *conspiracy number* $CN_n(v) = p_n(v) + d_n(v)$ is the smallest number of leaves (called conspirators) whose values must change for the minimax value of n to reach v. $CN_n(v) = 0$ iff $v = \text{MINIMAX}(n)$.

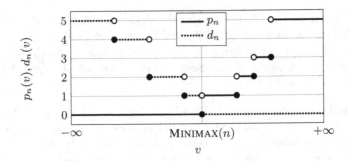

Fig. 1. Proof function p_n and disproof function d_n are monotonic staircase functions meeting at $(\text{MINIMAX}(n), 0)$.

2.1 Node Expansion

Our implementation of CNS follows PNS: iteratively select and expand a most proving node (mpn) and then update (dis)proof functions on the path to the root. So we define a CNS mpn.

Let $v_{root} = \text{MINIMAX}(root)$. Choose target values v_{max} for Max (the max player) and v_{min} for Min so that $v_{min} < v_{root} < v_{max}$. We discuss how in Sect. 2.2. We call $[v_{min}, v_{max}]$ *the search value interval* or search interval.[2] For fixed v_{max} and v_{min}, we say that Max (Min) *wins* if value v_{max} (v_{min}) is reached. To find a mpn we use $\text{SELECTMPN}(root)$, with pn $p_n(v_{max})$ and dn $d_n(v_{min})$ for every node n. Our CNS implementation — Algorithm 1 — differs from that of McAllester, as we alter both sides of the search interval at once.

Algorithm 1. Conspiracy number search

1: **function** CNS($root$)
2:     **while** not reached time limit **do**
3:         SETINTERVAL   ▷ Set v_{max} and v_{min}
4:         $n \leftarrow \text{SELECTMPN}(root)$
5:         Expand n and initiate new children by (1)
6:         Update nodes along path to the root using (2)
7: **function** SELECTMPN(n)
8:     **if** n is leaf **then**
9:         **return** n
10:     **else if** n is max-node **then**
11:         **return** SELECTMPN($\underset{s \in \text{children}(n)}{\text{argmin}} \; p_s(v_{max})$)
12:     **else**   ▷ n is min-node
13:         **return** SELECTMPN($\underset{s \in \text{children}(n)}{\text{argmin}} \; d_s(v_{min})$)

---

[2] This is the current likely range of the final root minimax value. It is analogous to the aspiration window of $\alpha\beta$ search.

2.2 Choosing the Search Interval

One way to pick the search interval is to set $v_{\max}$ and $v_{\min}$ a fixed difference from MINIMAX($root$), denoted v_{root},

$$
\begin{aligned}
v_{\max} &= v_{root} + \delta_p, \\
v_{\min} &= v_{root} - \delta_d,
\end{aligned}
\tag{3}
$$

where δ_p and δ_d are possibly equal constants. But it can help to modify the interval during search, e.g., by adjusting according to the root (dis)proof value,

$$
\begin{aligned}
v_{\max} &= \max_{v \in V}\{v : p_{root}(v) \leq P_{\max}\}, \\
v_{\min} &= \min_{v \in V}\{v : d_{root}(v) \leq D_{\max}\},
\end{aligned}
\tag{4}
$$

where $P_{\max}$ and $D_{\max}$ are possibly equal constants. This approach was used in the original CNS algorithm [13,16]. Search proceeds until the interval is sufficiently small, i.e., $v_{\max} - v_{\min} \leq \Delta$, where Δ is a constant indicating an acceptable error tolerance.

This method does not always converge, e.g., when the search is close to solving a position, or — if thresholds are too small — when the search stumbles into a stable position; in such cases it is better to increase thresholds and resume the search. Our approach below mixes (3) and (4). Notice that (5) generalizes (4).

$$
\begin{aligned}
v_{\max} &= \max_{v \in V}\{v : p_{root}(v) \leq \max(p_{root}(v_{root} + \delta_p), P_{\max})\}, \\
v_{\min} &= \min_{v \in V}\{v : d_{root}(v) \leq \max(d_{root}(v_{root} - \delta_d), D_{\max})\}.
\end{aligned}
\tag{5}
$$

To use CNS as a bot, we search until error tolerance is reached or time runs out and then pick the best move. Experiments show the best criterion for best move is the branch on which most time (leaf expansions in the subtree) is spent.

3 Sibling CNS

We convert a local heuristic — one that reliably scores relative strengths of siblings — into a global heuristic useful for our CNS player by adding relative errors, as follows. The evaluation of non-terminal game tree node n is given by

$$
\text{EVAL}(n) = \sum_{i=1}^{k} \sigma(p_{i-1}) \cdot e(p_{i-1} \to p_i),
\tag{6}
$$

where $p_0 \to p_1 \to \cdots \to p_k = n$ is the path from root p_0 to n, $\sigma(p_j) = 1(-1)$ if we (the opponent) are to move at p_j, and, for any child s of n, $e(n \to s) = \log \frac{E(n \to s_0)}{E(n \to s)}$ is the relative error at n with respect to s, where s_0 is a child of n with best score. We call this *siblings comparison evaluation function* (scef).

Generally, CNS constructs paths to terminal nodes, and then branches so that the player for whom the terminal node was losing tries to find another response in a subtree minimizing the cumulative error. So, the player tries to fall back on another most promising move of the entire tree.

Although SCNS — CNS with scef — explores good lines of play, the version we have described so far is wasteful, as CNS tends to expand all siblings whenever a new child is expanded. To avoid this, especially for unpromising children, we encode extra information in the (dis)proof function when creating a leaf. If a move has a high error compared to its best sibling, then to increase the minimax value of this move by this error will likely require many expansions. So, rather than initializing (dis)proof functions via a two-step staircase function (1), we use a multi-step staircase function, with the number of steps logarithmic in the difference between current and minimax values. Hence

$$p_n(v) = \begin{cases} 0 & \text{if } v \leq x \\ i & \text{if } i^\delta < 2^{(v-x)} \leq (i+1)^\delta \end{cases} \qquad d_n(v) = \begin{cases} 0 & \text{if } v \geq x \\ i & \text{if } i^\delta < 2^{(x-v)} \leq (i+1)^\delta \end{cases} \qquad (7)$$

where $x = \text{MINIMAX}(v)$, i is a positive integer and δ a positive rational. Using (7) to initialize non-terminal leafs, SCNS expands only siblings whose score diverges from that of the best sibling by at most δ. Depending on how values shift during search, other (weaker) siblings might be expanded if the minimax value changes by more than δ. With this modification, SCNS's search behavior is now closer to that of the human-like behaviour described above.

3.1   Gradual Forgetting of an Error

While cell energy works well in scef as a move's error estimate, it can assign a falsely high error to a good move. If SCNS spends much work at such a move (i.e., expands many nodes in the node's subtree) the initial error estimate should be corrected. We gradually decrease error as follows,

$$e'(n \to s) = e(n \to s) \cdot \max\left(1 - \frac{w_s}{W_{\max}}, 0\right), \qquad (8)$$

where w_s is work done at s, $W_{\max}$ is a constant parameter measuring the amount of work after which the error should be zero, and $e'(n \to s)$ is the adjusted error estimate.

3.2   Adding RAVE Statistics

One strength of MCTS Hex bots is their enhancement of move strength by the Rapid Action Value Estimate, an all-moves-as-first statistic [8]. So we added RAVE to SCNS. With each node we store a map from possible moves (cells) to the RAVE statistic, which consists of two integers: RAVE wins and losses. Statistics are updated whenever a terminal node is created by leaf expansion: for each node on the path from root to the node, we update RAVE values for each move played on the rest of the path.

Assume for the move $n \to s$ we have the RAVE win-loss statistic (w^R, l^R) of the player to move. Denote the number of RAVE games as $g^R = w^R + l^R$. We modify move error:

$$e'(n \to s) = (1 - \alpha) \cdot e(n \to s) + \alpha \cdot R_{\text{impact}} e^R(n \to s), \qquad (9)$$

where α indicates how quickly we shift into RAVE error

$$\alpha = \sqrt{\frac{g^R(n \rightarrow s)}{3R_{\text{shift}} + g^R(n \rightarrow s)}}, \tag{10}$$

$e^R(n \rightarrow s)$ is a move error computed by RAVE

$$e^R(n \rightarrow s) = \text{erf}^{-1}\left(\frac{l^R(n \rightarrow s) - w^R(n \rightarrow s)}{g^R(n \rightarrow s) + 1}\right), \tag{11}$$

erf^{-1} is inverse error function, and R_{shift} and R_{impact} are constant parameters which indicate how quickly we shift to RAVE error and the impact of RAVE error respectively.

RAVE encourages (discourages) moves that are more often involved in winning (losing) lines and gradually diminishes information from cell energy. SCNS often reaches terminal nodes, so RAVE values accumulate quickly. RAVE can be combined with gradual error forgetting by applying (8) on top of (9).

3.3   Transposition Table and Depth-First Implementation

PNS assumes (often incorrectly) that the complete tree can be stored in memory. The DFPNS algorithm overcomes this restriction via a depth-first implementation and transposition table [18]. A DFPNS enhancement — the $1+\varepsilon$ method — reduces the tendency of the search to jump around the tree [21]. The resulting algorithm is stronger than PNS and returns to the root only rarely [18,21].

We apply these three enhancements to CNS. Again, search rarely returns to the root, so updates to the search interval $[v_{\text{min}}, v_{\text{max}}]$ are infrequent. It may even happen that search stays too long in one subtree, in which case we want to force the search back to the root after a few expansions in order to refine the interval. A parameter for this is set according to the time-per-move setting.

3.4   Parallel SCNS

Our approach[3] is to mimic the parallelization of DFPN [20]: use a quick thread assignment that follows the natural CNS order, and halt thread execution once its task is redundant. This is achieved by using virtual wins and losses, and temporarily halting thread execution — returning the uncompleted portion of thread's task to the thread pool — once the thread has made MaxWorkPerJob recursive calls. So our parallel SCNS works as follows. See [20] for more details.

1. Replace (dis)proof numbers by (dis)proof functions: each operation — leaf initialization, node update, ... — is now done via (dis)proof functions.

---

[3] Another approach is to dynamically partition the CNS tree and evaluate subproblems in parallel. Lorenz achieved this for the restriction of CNS to 2 conpirators, i.e., effectively bounding proof function numbers at 2 [14].

2. Whenever search visits the root, set $v_{\max}$ and $v_{\min}$.
3. Navigate the search tree as in DFPNS, but with (dis)proof numbers $p_n(v_{\max})$ and $d_n(v_{\min})$ until search returns to the root.
4. Give each thread its own search interval, based on virtual (dis)proof functions.

4   Experimental Results

Using parallel SCNS, we implemented the Hex bot DEEPHEX on the Benzene framework [3]. Benzene includes virtual connection and cell energy computations, so as local SCNS heuristic we used the energy drop at each cell as described in Sect. 1.

We used two bots as opponents: WOLVE and MOHEX, each also implemented on Benzene. WOLVE uses $\alpha\beta$ Search with max-width pruning, with circuit resistance for heuristic. MOHEX — the strongest Hex bot since 2009 — uses MCTS with RAVE, patterns, prior knowledge estimation, progressive bias, and CLOP tuning of parameters [10]. WOLVE and MOHEX both compute virtual connections that prune moves and solve positions long before the game ends.

For openings, we used 36 relatively balanced single stone openings: a2 to k2, a10 to k10, b1 to j1, and b11 to j11.

We optimized parameters using CLOP (Sect. 4.1). Then we ran a knockout experiment to show feature importance (Sect. 4.2). Next we ran a tournament to show how strength increases with number of threads (Sect. 4.3). Finally, we ran a DEEPHEX vs. MOHEX tournament at competition settings (Sect. 4.4).

4.1   Parameter Optimization by CLOP

We optimized parameters using CLOP [6]. In the tuning process we played 30 s games, used MoHex as the reference opponent, and set the root-interlude (maximum number of node expansions before search must return to the root) to 20. The final parameter settings are based on 30 000 games; CLOP already found good settings after 20 000 games. DEEPHEX won 45 % of these CLOP-tuning games. Final settings are shown in Table 1.

The CLOP-tuned values hint at the effect of various parameters. δ measures the urgency of sibling expansion; 103 seems small, as moves become easily distinguishable with δ about 300. $P_{\max}$ and $D_{\max}$ are also small, so DEEPHEX prefers exploring promising lines deeply before diverging; a hand-tuned version of DEEP-HEX with $P_{\max} = D_{\max} = 1$ was strong, so we expected that the CLOP-tuned values to be close to 1; CLOP values 3,4 suggest that for DEEPHEX the best CNS behavior is not far from that of PNS. CLOP values show optional extension of the search interval by δ_p, δ_d is practically useless, as values 8,7 have negligible effect on performance. Surprisingly, RAVE impact is small. We guessed it would be important to incorporate the outcome of terminal nodes quickly, but values 211,782 show this is better done slowly. A similar conlusion holds for gradual error forgetting.

Table 1. Parameters tuned by CLOP for DEEPHEX.

Parameter	Value	Description
ε	0.41	ε tolerance
η	0.30	Allowed relative error of numbers in proof function
δ	103	The end of the first step in leaf initialization
$P_{\max}$	3	Proof threshold when setting $v_{\max}$
$D_{\max}$	4	Disproof threshold when setting $v_{\min}$
δ_p	8	Extending the search interval on max side
δ_d	7	Extending the search interval on min side
R_{shift}	211	RAVE error shift factor
R_{impact}	782	Impact of RAVE error
$W_{\max}$	1824	Error forgetting threshold

4.2 Knockout Experiment

Here we measure feature importance and accuracy of CLOP tuning. We tested many versions of DEEPHEX, each with either a feature off or a parameter slightly changed. For each version we played 720 matches against MOHEX (10 times for each opening) at 30 s/move and then — to measure scaling — at 60 s/move. See Table 2.

Table 2. Knockout experiment results showing win percentage over MOHEX by differ settings in DEEPHEX. The DEEPHEX versions are: (base) all features, all parameters with CLOP settings, (a) basic leaf initialization, (b) $1 + \varepsilon$ method off, (c) exact proof functions (approximation off), (d) gradual error forgetting off, (e) RAVE off, (f) gradual error forgetting and RAVE both off, (g) smallest possible thresholds inducing smaller search interval, (h) as in (g) but extending search interval to at least 100 on each side, (i) larger thresholds for setting the search interval.

id	version	30 s	60 s
(base)	CLOP tuned	60.0	52.6
(a)	2-step (dis)proof function in leafs	20.3	23.1
(b)	no $1 + \varepsilon$ method ($\varepsilon = 0$)	57.5	50.6
(c)	Exact proof functions ($\eta = 0$)	56.1	48.9
(d)	no gradual error forgetting	56.7	51.9
(e)	no rave	51.9	57.5
(f)	pure scef	52.4	55.6
(g)	$P_{\max} = D_{\max} = 1$	52.1	52.5
(h)	$P_{\max} = D_{\max} = 1$ and $\delta_p = \delta_d = 50$	59.4	51.5
(i)	$P_{\max} = D_{\max} = 5$	56.0	53.9

As expected, at 30 s/move the CLOP-tuned version is strongest. The most critical feature is better leaf initialization via the multi-step proof function. RAVE is beneficial at 30 s/move but less so at 60s/move; perhaps DEEPHEX tuning parameters (or at least the settings found by CLOP) are sensitive to time/move.

4.3   Multi-threaded Tournament

Here we show how program strength scales with number of threads. 13 bots competed: 1,2,4,8-thread DEEPHEX; 1,2,4,8-thread MOHEX; 1,3,7-thread MOHEX plus 1 thread for solver; 1-thread WOLVE; 1-thread WOLVE plus 1 thread for solver. In each game each bot had 30 s/move. Each bot played each other bot two times on each opening, once as black (1st-player) and once as white (2nd-player). So each bot played 864 of the 5616 tournament games.

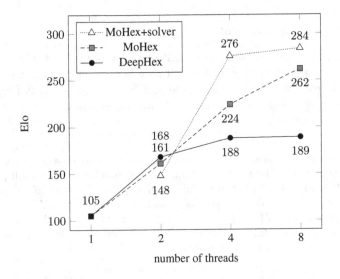

Fig. 2. Tournament results. Each point has error within 14 Elo, with 80 % confidence. Reference player is 1-thread WOLVE, BayesElo score 0.

Figure 2 shows tournament results. Scores are BayesElo [5] with respect to reference player WOLVE (win rate .31, score 0); WOLVE and Wolve+solver (score 23) are not shown. MOHEX scales well up to the maximum 8 threads; this is perhaps not surprising, as MCTS strength typically increases uniformly with number of simulations and parallelizes relatively easily. MOHEX+solver scales well up to 4 threads, but is only slightly stronger at 8 threads. The latter is perhaps because, with solver effectively taking over end games, the difference in opening play between 3-thread MOHEX and 7-thread MOHEX is not enough to change many outcomes. DEEPHEX scales as well as MOHEX up to 2 threads, but then more poorly. This drop in scaling efficiency is more pronounced than a similar drop in scaling efficiency of parallel DFPN [20], perhaps because the underlying parallelization technique — prevent thread convergence in the search tree via virtual wins and losses — works better in PNS than in CNS. More work is needed to explore this behavior.

4.4 DeepHex versus MoHex

Here we simulated a competition tournament. Each bot had 12 threads. MoHEX used its strongest settings: 1 thread for its DFPNS solver and 11 for MCTS. DEEPHEX does not yet have game-length time control, so to approximate 30 m/game, we allowed it 90 s/move, as, for almost every game, each bot knows the winner before 40 moves.

We played 9 rounds, each with 72 games (2 per opening, each bot once black and once white), a total of 648 games. DEEPHEX had a .448 win rate. Earlier experiments suggest that more tuning might strengthen DeepHex, for instance tuning parameters by CLOP for larger time limits or for more threads.

Under these settings MoHEX seems stronger in early play but DEEPHEX sees further in complicated positions. Figure 3 shows a typical game, where MoHEX pushes DeepHex into a losing position before DEEPHEX escapes.

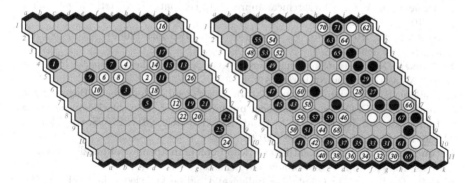

Fig. 3. DEEPHEX (Black) escapes against MoHEX. After 26.j5 DEEPHEX sees its loss with PV i6 h6 i5 i10 i9 h10 h9 g10 g9 f10 g8 h2 k1 e9 c9 d8 a8 d7 d9 e8 c7 d6 a7 b5. But MoHEX sees neither this nor its win after 28.h6 and blunders with 30.h11 instead of i10. After 60 s of search DeepHex sees a win 31.h10 with PV g11 g10 f11 f10 e11 e10 d11 d10 c11 b10 c10 c7 d9 b7 f8 b6 d3 d4 d1 b9 f2 f3 g2 g3 c8 b8 h2 b2 b3 a3 b5 c4 b4 c3. With 34.f11 the MoHEX search threads score .62 before the solver thread finds the loss.

In this tournament the average game length for a MoHEX (DEEPHEX) win is 48.6 (61.2) moves. MoHex wins almost all short games, DEEPHEX wins almost all long ones. See Table 3. MoHEX seems strategically stronger, often — perhaps because it is ahead — making simplifying moves. DEEPHEX seems tactically further-sighted, often — perhaps because it is behind — making complicated moves. A research challenge is to mix these two behaviors.

Table 3. DEEPHEX win rate by game length.

Length	26–40	41–50	51–60	61–70	71–80	81–93
Rate	.04	.20	.52	.73	.83	1.00

5   Conclusions and Further Research

We showed the strength of Sibling Conspiracy Number Search features by competing our SCNS Hex bot with MCTS bot MoHEX and $\alpha\beta$ bot WOLVE. By far the most critical feature is to initialize leaf (dis)proof functions via a multi-step— rather than 2-step— staircase function. Also, we showed a parallel version of SCNS. Our parallel SCNS Hex bot scales well — as well as MoHEX — with 2 threads, but less efficiently with 4 or 8 threads. An open problem is to parallelize SCNS more effectively.

References

1. Allis, L.V.: Searching for Solutions in Games and Artificial Intelligence. PhD thesis, University of Limburg, Maastricht, The Netherlands (1994)
2. Anshelevich, V.V.: A hierarchical approach to computer Hex. Artif. Intell. **134**(1–2), 101–120 (2002)
3. Arneson, B., Henderson, P., Hayward, R.B.: Benzene (2009). http://benzene.sourceforge.net/
4. Breuker, D.M.: Memory versus Search in Games. PhD thesis, Maastricht University, Maastricht, The Netherlands (1998)
5. Coulom, R.: Bayesian elo rating (2010). http://remi.coulom.free.fr/Bayesian-Elo
6. Coulom, R.: CLOP: confident local optimization for noisy black-box parameter tuning. In: van den Herik, H.J., Plaat, A. (eds.) ACG 2011. LNCS, vol. 7168, pp. 146–157. Springer, Heidelberg (2012)
7. Gardner, M.: The 2nd scientific american book of mathematical puzzles and diversions, Chap. 7, pp. 78–88. Simon and Schuster, New York (1961)
8. Gelly, S., Silver, D.: Combining online and offline knowledge in UCT. In: 24th ACM ICML, pp. 273–280 (2007)
9. Henderson, P.: Playing and solving Hex. PhD thesis, UAlberta (2010). http://webdocs.cs.ualberta.ca/~hayward/theses/ph.pdf
10. Huang, S.-C., Arneson, B., Hayward, R.B., Müller, M., Pawlewicz, J.: MoHex 2.0: a pattern-based MCTS Hex player. In: van den Herik, H.J., Iida, H., Plaat, A. (eds.) CG 2013. LNCS, vol. 8427, pp. 60–71. Springer, Heidelberg (2014)
11. Iida, H., Sakuta, M., Rollason, J.: Computer shogi. Artif. Intell. **134**(1–2), 121–144 (2002)
12. Kishimoto, A., Winands, M., Müller, M., Saito, J.-T.: Game-tree searching with proof numbers: the first twenty years. ICGA J. **35**(3), 131–156 (2012)
13. Klingbeil, N., Schaeffer, J.: Empirical results with conspiracy numbers. Comput. Intell. **6**, 1–11 (1990)
14. Lorenz, U.: Parallel controlled conspiracy number search. In: Monien, B., Feldmann, R.L. (eds.) Euro-Par 2002. LNCS, vol. 2400, pp. 420–430. Springer, Heidelberg (2002)

15. Lorenz, U., Rottmann, V., Feldman, R., Mysliwietz, P.: Controlled conspiracy number search. ICCA J. **18**(3), 135–147 (1995)
16. McAllester, D.: Conspiracy numbers for min-max search. Artif. Intell. **35**(3), 287–310 (1988)
17. McAllester, D., Yuret, D.: Alpha-beta conspiracy search. ICGA **25**(1), 16–35 (2002)
18. Nagai, A.: Df-pn Algorithm for Searching AND/OR Trees and Its Applications. Ph.d. thesis, Department of Information Science, University Tokyo, Tokyo, Japan (2002)
19. Pawlewicz, J., Hayward, R.: Sibling conspiracy number search. manuscript (2015)
20. Pawlewicz, J., Hayward, R.B.: Scalable parallel DFPN search. In: van den Herik, H.J., Iida, H., Plaat, A. (eds.) CG 2013. LNCS, vol. 8427, pp. 138–150. Springer, Heidelberg (2014)
21. Pawlewicz, J., Lew, L.: Improving depth-first PN-search: 1 + ϵ trick. In: van den Herik, H.J., Ciancarini, P., Donkers, H.H.L.M.J. (eds.) CG 2006. LNCS, vol. 4630, pp. 160–171. Springer, Heidelberg (2007)
22. Saito, J.-T., Chaslot, G.M.J.-B., Uiterwijk, J.W.H.M., van den Herik, H.J.: Monte-Carlo proof-number search for computer Go. In: van den Herik, H.J., Ciancarini, P., Donkers, H.H.L.M.J. (eds.) CG 2006. LNCS, vol. 4630, pp. 50–61. Springer, Heidelberg (2007)
23. Schaeffer, J.: Conspiracy numbers. Artif. Intell. **43**(1), 67–84 (1990)
24. van der Meulen, M.: Parallel conspiracy-number search. Master's thesis, Vrije Universiteit Amsterdam, The Netherlands (1988)
25. Winands, M.: Informed Search in Complex Games. PhD thesis, Universiteit Maastricht, Maastricht, The Netherlands (2004)
26. Winands, M.H.M., Schadd, M.P.D.: Evaluation-function based proof-number search. In: van den Herik, H.J., Iida, H., Plaat, A. (eds.) CG 2010. LNCS, vol. 6515, pp. 23–35. Springer, Heidelberg (2011)

Parameter-Free Tree Style Pipeline
in Asynchronous Parallel Game-Tree Search

Shu Yokoyama[1], Tomoyuki Kaneko[1]([✉]), and Tetsuro Tanaka[2]

[1] Graduate School of Arts and Sciences, The University of Tokyo, Tokyo, Japan
kaneko@graco.c.u-tokyo.ac.jp
[2] Information Technology Center, The University of Tokyo, Tokyo, Japan

Abstract. Asynchronous parallel game-tree search methods are effective in improving the playing strength by using many computers connected through relatively slow networks. In game-position paralleliza- tion, the master program manages a game-tree and distributes positions in the tree to workers. Then, each worker asynchronously searches the best move and the corresponding evaluation for its assigned position. We present a new method for constructing an appropriate master tree that provides more important moves with more workers on their sub- trees to improve the playing strength. Our contribution introduces two advantages: (1) being parameter free in that users do not need to tune parameters through trial and error, and (2) efficiency suitable even for short-time matches, such as one second per move. We implemented our method in chess with a top-level chess program STOCKFISH and evalu- ated the playing strength through self-plays. We confirm that the playing strength improves with up to sixty workers.

1 Introduction

Parallelization of game-tree search has been extensively studied to improve the game programs' playing strength, especially in chess and its variants. Several effi- cient methods have been developed in hardware parallelization [2,4], in thread- level parallelization [14], and for tightly connected computers [1,5,12,18].

Recently, as grid computing has become popular, new approaches utilizing computational resources placed in different locations connected through wide- area networks have been proposed. Game position parallelization (GPP) [16] is one such method actually showing steady improvements in playing strength. In that method, a local computing unit (we call it a *worker*, even though it could be a cluster of tightly connected computing nodes) assumes a position assigned to it. Each worker runs its own game-tree search independently, and then, the master integrates the workers' results to reach a decision.

This study presents pipeline GPP (P-GPP), which extends both Optimistic Pondering [8] and GPP, by improving worker management. In P-GPP, positions

T. Kaneko—A part of this work was supported by JSPS KAKENHI Grant Number 25330432.

A. Plaat et al. (Eds.): ACG 2015, LNCS 9525, pp. 210–222, 2015.
DOI: 10.1007/978-3-319-27992-3_19

are assigned to workers on the basis of realization probabilities [17] automatically acquired from game records and a playing program. This automation frees users from the need to tune heuristic parameters, whereas existing methods need many configuration parameters. Experiments demonstrate P-GPP's effectiveness with up to sixty workers, with results comparable to those shown in the literature [8]. Therefore, P-GPP deserves further study of both its effectiveness and usability in terms of being parameter free.

The remainder of this paper is organized as follows. The next section reviews related research. The third section introduces the GPP framework, and the fourth section presents the P-GPP method. The fifth section presents our experimental results in chess and improvements in playing strength through self-play. The last section provides our concluding remarks.

2   Related Work

In this section we discuss parallelisation of alpha-beta pruning (Sect. 2.1) and integration of computing resources through the Internet (Sect. 2.2).

2.1   Parallelization of Alpha–Beta Pruning

State-of-the-art sequential algorithms on game-tree search have been built upon alpha–beta pruning [13] with many enhancements. Basically, playing strength improves when a program searches more deeply, assuming that adequate evaluation functions have been provided. Therefore, various parallel search algorithms have been developed to improve strength by exploring game trees in a shorter time by using more processors. The best solution depends on users' environments, because there is a well-known trade-off in the design of parallelization; an increase in shared information among processors increases pruning effectiveness, while communication to share information inevitably incurs overheads that degrade efficiency. State-of-the-art sequential algorithms prune branches aggressively (e.g., by late move reductions[1]) by utilizing information of the tree explored already, e.g., by transposition tables, $\alpha\beta$-windows, killer moves, and history tables.

In parallel search methods in shared memory environments (e.g., Principal Variation Splitting [14], Dynamic Tree Splitting[2]), transposition tables are naturally shared. However, in effective parallelization in distributed environments, only a part of a transposition table is shared [3,8] because the cost for full sharing is not beneficial overall. Still, in major approaches including YBWC and its enhancements [5,18], APHID [1], and TDSAB [12], $\alpha\beta$-windows or equivalent information are shared in frequent communication. Therefore, they work more effectively in a network with higher quality, e.g., Infiniband, than with an ordinary one. Also, many practical systems incorporated hybrid parallelization including hardware (e.g., DEEPBLUE [2] and HYDRA [4]).

---

[1] http://www.glaurungchess.com/lmr.html (Last access: February 2015).
[2] https://www.cis.uab.edu/hyatt/search.html (Last access: February 2015).

2.2   Integration of Computing Resources Through the Internet

Recently, accessing computational resources placed elsewhere has become easy, if one permits relatively high latency and limited bandwidth to reach them, e.g., through ordinary Ethernet or wide-area networks. Hence, several new methods for utilizing such resources have been developed to improve playing strength further. Owing to network limitations, these methods are designed to work with little inter-node communication. For example, majority voting requires only communication regarding a position to search and vote for a move [15].

In Optimistic Pondering [7–9] and GPP [16] which were developed independently, workers are assigned distinct positions and search independently without sharing information. In Optimistic Pondering, the goal is to increase "ponder-hit" rate as well as to begin pondering as many plies earlier as possible. Pondering gains additional thinking time to deepen the search when pondering hits, i.e., the position assigned to a worker is actually realized in the game. In GPP, the goal is to conduct minimax search cooperatively by integrating search trees explored by hundreds of workers [16].

A notable advantage of these approaches is that they lend themselves well to combination with existing parallelization methods. The effectiveness of Optimistic Pondering in integrating workers running in YBWC mode was demonstrated in GRIDCHESS [8]. A combination of majority voting, GPP, and shared memory parallelization on each worker is used for playing shogi in Akara [10].

3   Game Position Parallelization

This section introduces the details of Game Position Parallelization (GPP), on which our work, P-GPP, is constructed (Sect. 3.1). GPP is based on a master/worker model, with a typical worker being a universal chess interface (UCI)[3] chess engine (Sect. 3.2). Finally, a comparison is made with similar systems (Sect. 3.3).

3.1   Master Tree

GPP conducts minimax search by integrating the results obtained locally by workers. The master constructs a *master tree* for task assignments for workers. The root of a master tree corresponds to the current position, and the number of nodes of the master tree must be the number of workers available. Assume that we have six equivalent workers A, B, C, D, E, and F. The master constructs a game tree rooted at the current position, which has five leaves. Figure 1 shows an example of a master tree. A node depicted in a rounded rectangle corresponds to a position similar to those in usual game trees. A leaf "others" enclosed in a rectangle is a special position, in which the position is the same as that of the parent but for which moves to be searched are limited, excluding moves already

---

[3] http://www.shredderchess.com/chess-info/features/uci-universal-chess-interface. html (Last access: February 2015).

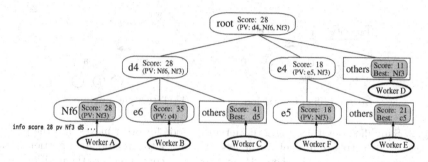

Fig. 1. A master tree with the initial position at the root. Each leaf has a worker assigned. For an internal node "d4", workers A, B, and C work on it cooperatively. Workers A and B assume responsibility for two child positions that are assumed to be the best ("Nf6") and second best ("e6") successors, and C assumes responsibility for the remaining moves. The best move and its score with respect to the max player are kept at each leaf shown in a gray rounded rectangle. The scores and best moves of internal nodes are computed, as in minimax search.

covered by its siblings. For example, at position "d4" in the figure, the worker C does not search moves Nf6 and e6 because workers A and B are working on them, respectively. Therefore, at each internal node, the children cover all moves without duplication.

Each worker then independently execute game-tree search for a node and periodically reports the best move (more precisely, a sequence of best moves, the principal variation, PV) with its evaluation score. The master then integrates the best moves and scores. Communication required in this process is supported by standard game protocols, UCI for chess and universal shogi interface (USI)[4] for shogi. For example, the command go searchmoves restricts the moves searched at an "others" leaf. The use of a text protocol improves software modularity.

GPP aims to achieve steady improvements in playing strength with hundreds of workers in slow networks [16]. Therefore, neither transposition tables nor $\alpha\beta$ windows are shared. If the network quality is sufficient for them to be shared, the methods introduced in Sect. 2.1 are preferable to GPP.

3.2   Tree Growth and Tree Style Pipeline

To improve playing strength in GPP, a master tree must be carefully constructed such that a more relevant move should have more workers than less impor-tant moves. The problem here is the difficulty in identifying relevant moves in advance.

Having the master tree of the previous position available while playing a game provides two advantages. First, relevant branches can be estimated from the information stored in the tree. Second, workers working on common nodes between the previous and new trees can continue searching without interruption.

---

[4] http://www.glaurungchess.com/shogi/usi.html (Last access: February 2015).

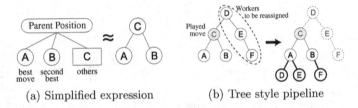

(a) Simplified expression   (b) Tree style pipeline

Fig. 2. Simplified notation (Fig. 2a): Merging a node for other moves into the parent, makes the right tree equivalent to the left one. Either tree represents a portion of the master tree shown in Fig. 1, where "Parent Position" corresponds to the position "d4". Tree-style pipeline (Fig. 2b): The left tree is the simplified notation for Fig. 1. When the board situation changes, the master tree grows. Workers working on positions that are not descendants of the new root (D, E, and F) are collected and assigned to the newly created leaves. (See also Sect. 4.)

In this study, we call the idea related to these advantages *tree style pipeline*. The idea has already been adopted in two different methods: tree of pondering pipeline in Optimistic Pondering [9], and a recent GPP method [11].

In tree style pipeline, nodes in the previous master trees that remain effective in the new tree are preserved. The remaining nodes are discarded, and their corresponding workers are assigned to new expanded leaves, as shown in Fig. 2b. As a simple illustration, note that we draw a tree merging an "others" leaf to its parent, as explained in Fig. 2a. An advantage in tree style pipeline is that a worker on a shaded node in Fig. 2b continues searching without interruption in the transition from the previous tree to the new one. In GPP, a normal leaf sometimes becomes an "others" leaf, when its child is expanded. Node A in Fig. 2b is in such a situation. A worker working on such a node is stopped and immediately restarted with restriction of moves already expanded (i.e., D and E in this example). This restart's negative effect must be rather limited because the contents in a transposition table are preserved in each worker. Workers outside the common tree are collected and used to grow relevant branches of the tree. In expanding a leaf of the tree, the current best move at the leaf is a promising candidate to expand. However, it is still not clear which leaves (including "others" nodes) are to be expanded and how many new workers each leaf is worth. This study's primary contribution is to present a new criterion and a procedure for this problem.

3.3 Comparison with Similar Systems

Table 1 summarizes related work similar to our work. GridChess combines Optimistic Pondering and tree style pipeline. The main difference between that work and ours is minimax integration. Optimistic Pondering focuses on pondering and does not perform minimax backup in a master tree. Consequently, tasks assigned to workers are not disjoint in Optimistic Pondering, i.e., it does not have any "others" leaf in GPP, such as worker C in Fig. 2a. Lacking minimax

Table 1. Comparison of distributed search methods and systems similar to GPP. The first column is the name of the method or system. The second (third) column indicates whether minimax backup (tree style pipeline) is used. The fourth column describes when new leaves in the master tree are expanded. "New root" means when the root of the master tree changes. The fifth column describes what information source is used to select leaves to be expanded.

	Minimax	Pipeline	Growth	Source
GRIDCHESS [8] (chess)	-	Yes	Anytime	PV
GPP [16] (shogi)	Yes	(Yes)	New root	Shallow search
P-GPP(chess)	Yes	Yes	New root	Previous PV+Hash

backup can create a problem. Suppose that a move at the root seems promising within a certain search depth, but actually is a blunder that deeper search can reveal. Through constructing a larger master tree by investing more workers, GPP might detect it earlier than a single worker does. However, Optimistic Pondering cannot see it until the search of the root worker reaches a sufficient depth, even if many additional workers are involved. Additionally, Optimistic Pondering changes its master tree more dynamically. While frequent changes might improve playing strength, many heuristic parameters need to be tuned, regarding when to start and stop pondering a position.

The idea of GPP was first developed in connection with shogi, integrating more than 300 ordinary computers and reported in the literature [16]. The original version does not have tree style pipeline. Instead, a shallow search is used to determine the moves to be expanded at each step in the construction of a master tree. Because this shallow search decreases the available time for the primary search, our work introduced an alternative method for growing a master tree. Additionally, heuristic parameters r_0 and r were used in the assignment of workers; for the n-th promising move at a position having N workers, $r_0 r^{n-1} N$ workers are assigned. The values were $r_0 = 1/4$, $r = 3/4$ for the root, and $r_0 = r = 1/2$, otherwise. Later, many heuristic parameters including domain-dependent ones were introduced, when it was used in human–computer shogi matches [10,11]. Therefore, our work is the first method incorporating both GPP and tree style pipeline, simultaneously, eliminating heuristic parameters.

4   P-GPP: Pipeline GPP with Parameter-Free Approach

This section presents our method (P-GPP) extending GPP. When the board is changed by a move played by a program or its opponent, the master tree for GPP must be updated so that the node corresponding to the new position becomes the root in a new tree. P-GPP constructs its new tree using the following three steps, to find the best tree with respect to playing strength.

1. Unreachable nodes from the new root position are removed from the tree, and the corresponding workers are collected (e.g., node D, E, and F in Fig. 2b).

2. The greedy algorithm presented here determines the number of workers for each node (except for an "others" leaf), using realization probabilities.
3. For each node l and the number of workers n identified in the previous step, a concrete sub-tree rooted at l having n nodes is created, considering the transposition table of the worker at l.

The main contributions of this study are the new methods for steps 2 and 3 presented in Subsects. 4.1 and 4.2. For initiating the pipeline process, we define the initial master tree as a tree having only its root as the starting position. Alternatively, opening books can be used. In Sect. 4.3, we discuss adaptation of realization probability, that is essential to implement step 2 and 3 in practice.

4.1   Utility of Master Tree Based on Realization Probability

We introduce the method assuming that the realization probability is available for all nodes. The utility $U(T)$ of a master tree T is the summation of the depth weighted by its realization probability:

$$U(T) = \sum_{v \in V(T)} p_v d_v, \tag{1}$$

where $V(T)$ is the set of vertices in the tree T, d_v is the depth of node v and p_v is its realization probability. GPP works effectively, if a position included in the master tree will be reached in an actual game in the future, and it is more beneficial when the position is further from the root to have more thinking time before it manifests. So, if we can predict the future, then narrow, deep trees are preferable for that purpose. However, when we miss a prediction, many workers must be reallocated, and their results do not contribute to the playing. This means that the total number of workers collected in the future must be minimized. Thus, the expectation over the probability in Eq. (1) is adopted.

The realization probability of a node, defined as the product of the transition probability of each move [17], is the probability that the corresponding sequence of moves is actually played. By definition, the realization probability of the root is one. We also assume that the summation of the transition probability of all legal moves is one in each position. We ignored such edges in counting the depth in computing the utility that represent "others" moves, because they do not reduce the search space compared to normal edges. In our simplified notation of a tree (see Fig. 2a), each internal node corresponds to a leaf led by all other moves. Hence, the summation of the realization probabilities for all the nodes in a simplified tree is always one.

4.2   Greedy Growth Algorithm

Assuming that the realization probability of each node is available, we present a simple greedy algorithm based on iteratively adding a node having the largest realization probability. Figure 3 shows a step of the greedy algorithm, where

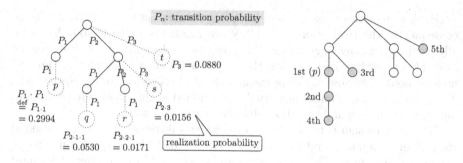

Fig. 3. Example of greedy steps: for a tree drawn with a solid line with candidates, each of which is a dotted circle with its probability, node p having the largest probability is added. After the fifth step, the tree in the right figure is obtained.

P_i is the transition probability of a rank i move, and the values are from our experiments listed in Table 2. This algorithm yields a tree of maximum utility in Eq. (1).

The proof is based on the change in the utility when a node is added (and symmetrically removed). Let T_1 be a tree, x a leaf in T_1, and T_0 the tree with x removed from T_1. The differences between T_0 and T_1 are node x which exists only in T_1 with probability p_x and parent x', which exists both in T_0 and T_1 but $p_{x'}$ in T_0 is greater than that in T_1 by p_x. Thus, for the probability p_x, the depth of x is added to $U(T_1)$ while the depth of x' is added to $U(T_0)$, where the difference in the depth of x and x' is 1. Therefore, we have $U(T_1) - U(T_0) = p_x$.

Here is a sketch of the proof by contradiction. Let T be the tree yielded by our greedy algorithm. We assume that there exists a tree T' having the maximum utility that is strictly better than that of T, i.e., $U(T') > U(T)$, $T \neq T'$ and $|V(T)| = |V(T')|$. Then, we select two nodes, v^* from $T \setminus T'$ and v from $T' \setminus T$. Let $v^* \in T \setminus T'$ be the node added at the earliest step. There exists a node v in $T' \setminus T$, such that $p_{v^*} > p_v$.[5] Thus, when we generate T'' by replacing node v in T' with node v^*, we have $U(T'') > U(T')$. This contradicts the assumption that T' has the maximum utility. $\square$

4.3 Realization Probability in Practice

For the realization probability in P-GPP, we used the empirical transition probability that expert players play a move of n-th rank when moves are ranked by a worker program. While the class of a move (e.g., check) is used in the original literature [17], we believe the ranks determined by the scores in the previous master tree are more reliable sources here. The probabilities regarding the rank, the only parameters that P-GPP requires, are obtained from a worker program and game records, as shown in Table 2 in the experiments discussed later.

---

[5] By the definition of the greedy algorithm, we have $p_{v^*} \geq p_v$ for any $v \in T' \setminus T$. In the special case that $p_{v^*} = p_v$ for all $v \in T' \setminus T$, utility $U(T')$ equals $U(T)$ by the definition of v^*, and this contradicts the assumption that $U(T') > U(T)$.

Having the score in each node, including "others" in the tree, the rank and consequently the realization probability are identified for each node by virtually sorting nodes by their scores. However, the ranks and probabilities are not available for newly added nodes at this step. Therefore, we virtually add an n-th node as needed without knowing the actual n-th best move leading to the node. Then, we count the number of virtual nodes added in this process for each node and recover the original tree discarding these virtual nodes.

Now the problem is to construct an effective subtree with n nodes rooted at the leaf, given a node l and the number n to be added. If $n = 1$, it is sufficient to expand the best move. Otherwise, this can be achieved by extracting the n most valuable positions from the transposition table of the corresponding worker (if l is a leaf) or the worker of its "others" leaf (if l is an internal node). We used the depth searched under a position for the criterion in this selection. These positions are sent from each leaf to the master every time the new master tree is constructed. Because the estimated data size is about one kilobyte for 32 entries, it can be expected to consume negligible time and network resources.

5   Experiments

To evaluate P-GPP, we implemented our method in chess. We first show the relative frequency (or empirical probability) of moves with respect to their rank, obtained from game records. Then, by using the frequency as the transition probability, self-play experiments are conducted.

5.1   Configurations

We adopted STOCKFISH DD[6] as a worker program, because it is an open source program and is expected to be one of the strongest chess programs. We added the function of reporting information described in Sect. 4.3, extending the UCI protocol. Each worker and the master are connected via standard TCP sockets. The master is implemented in C++ with the boost/asio library. For a worker, a utility program *netcat* is adopted as a proxy connecting stdin/stdout and a TCP socket. To simulate a distributed environment, we used at most 64 cores in two computers each of which is equipped with two Intel Xeon E5-4650 processors. STOCKFISH ran as a sequential program using a single thread. Each worker was allowed to use 32MiB (STOCKFISH uses 16 bytes per position) for its transposition table.

5.2   Empirical Probability with Respect to the Rank of a Move

Table 2 shows the frequency of moves played for each rank. Twelve game records played between DEEPBLUE and Kasparov consisting of 1 091 plies were used. For each position, all moves are scored using fifteen-depth search and sorted to

---

[6] https://s3.amazonaws.com/stockfish/stockfish-dd-src.zip.

Table 2. Frequency of move w.r.t. the rank, evaluated by STOCKFISH

Rank	1	2	3	4	5	6	7	8	9	10	11+
Frequency	0.5472	0.1769	0.0880	0.0522	0.0293	0.0247	0.0211	0.0128	0.0082	0.0110	0.0284

get the rank. We classified the ranks in eleven classes, from first to tenth, and the eleventh or greater. The result shows 54.7 % for the first-ranked move and 81 % in the top three moves.

5.3 Improvements in Strength

We conducted self-play experiments and showed the winning probability of several variations of the presented system against a sequential program. The sequential program is nearly the same as the original STOCKFISH, except that it ponders the current position instead of a future predicted position. The reason for the adjustment is to average the effect of ponder-hits and misses. Additionally, the sequential program is nearly the same as the presented method with a single worker. A program XBOARD managed matches in judging and recording the results and a program POLYGLOT[7] is used to connect XBOARD and STOCKFISH. The book used was **performance.bin**.[8] The opening was randomly chosen from the book by POLYGLOT. The win rate here is defined as the probability of wins plus half of the probability of draws, following the literature [6].

To consider the communication overhead in distributed environments, we imposed a thinking-time penalty on the proposed program. While the sequential program was given $1\,000$ ms to think per ply, the presented system P-GPP was given only 950 ms. We believe that 50 ms is more than sufficient for the communication in the presented method. In addition to P-GPP, we measured the win rate of "Linear Speedup" and "Random Growth." The former is the sequential program given n-times thinking time instead of using n workers.[9] This program gives the upper bound of the win rates gained by the ideal parallelization without overhead. The latter is a variation of P-GPP, ignoring transposition tables; it adds the position after the best move and randomly picked up $n-1$ positions, when adding n nodes to a leaf at the step in Sect. 4.3. For each configuration, $1\,000$ games were played alternating black and white.

Figure 4a shows the win rate of the parallel programs against the sequential one. The horizontal axis indicates the concurrency in the log-scale, and the error bars indicate as 95 % confidence interval. In P-GPP, the win rate increases with the number of workers. With a single worker, the win rate was 48.1 %, not even reaching 50 %. This was apparently caused by the thinking-time penalty. With 32 and 60 workers, our method achieved 62.5 % and 64.6 % win rates, respectively. The improvements did not reach those in the linear speedup, but they are similar to those reported in Optimistic Pondering built upon the cluster TOGA [8].

---

[7] Version 1.4w29 http://www.geenvis.net/polyglot1.4w29.zip.

[8] http://wbec-ridderkerk.nl/html/downloada/lacrosse/performance.rar.

[9] The opponent was configured not to use this additional time in pondering.

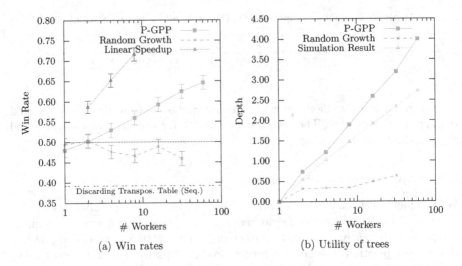

(a) Win rates   (b) Utility of trees

Fig. 4. Left (Fig. 4a): win rates of parallel systems in self-play. Right (Fig. 4b): observed utility. The average number of played plies included in the master trees, in self-play for the player's turn, with a "Simulation Result" that is $U(T)$ of the trees built virtually using only the greedy algorithm with the realization probability.

Therefore, we may conclude that our method scaled reasonably at least up to 60 workers. In contrast, "Random Growth" did not improve the playing strength. In the case of a single worker, the win rate was even, as the two players are the same. However, it became weaker with four or more workers. These results show the importance of master trees. When the master tree is random, only the worker working on the root contributes to playing strength. Moreover, the transposition tables of the workers tend to be filled with irrelevant positions, possibly degrading strength. To examine this result further, we measured the win rate of such a sequential program, which clears its transposition table every time it plays a move. The win rate was only 39.2 %, explaining the contribution of the transposition table to strength.

Figure 4b shows the utilities of the master trees constructed in self-play experiments, with simulated utilities considering only the realization probabilities listed in Table 2. The observed utilities are the average maximum depth of a position reached in actual games. In P-GPP, the utilities increase along with the concurrency and go beyond the simulated results, most likely because there are sometimes fewer legal moves in some positions than the branching factor of the simulated tree, and it is easier for STOCKFISH to predict moves played by the same program than to predict moves played by Kasparov's or by DEEPBLUE. The utilities did not reach 1.0 in "Random Growth."

6   Conclusion

We demonstrated that P-GPP, a new asynchronous parallel game-tree search method, works effectively in chess. P-GPP has two advantages: it is parameter-free in that users do not need to tune parameters through trial and error, and it is suitably efficient even for short-time matches. We confirmed that playing strength improves with up to sixty workers. The win rates are comparable to those of an existing method [8]. Therefore, we believe that P-GPP is a simple and promising alternative to existing methods. Interesting future work would involve scalability up to hundreds of workers.

References

1. Brockington, M.: Asynchronous Parallel Game-Tree Search. Ph.D. thesis, University of Alberta (1998)
2. Campbell, M., Hoane Jr., A.J., Hsu, F.H.: Deep blue. Artif. Intell. **134**(1–2), 57–83 (2002)
3. Donninger, C., Kure, A., Lorenz, U.: Parallel brutus: the first distributed, FPGA accelerated chess program. In: Proceedings of the 18th International Symposium on Parallel and Distributed Processing, 2004, p. 44, April 2004
4. Donninger, C., Lorenz, U.: The chess monster hydra. In: Becker, J., Platzner, M., Vernalde, S. (eds.) FPL 2004. LNCS, vol. 3203, pp. 927–932. Springer, Heidelberg (2004)
5. Feldmann, R.: Game Tree Search on Massively Parallel Systems. Ph.D. thesis, University of Paderborn (1993)
6. Heinz, E.A.: New self-play results in computer chess. In: Marsland, T., Frank, I. (eds.) CG 2001. LNCS, vol. 2063, pp. 262–276. Springer, Heidelberg (2002)
7. Himstedt, K.: An optimistic pondering approach for asynchronous distributed game-tree search. ICGA J. **28**(2), 77–90 (2005)
8. Himstedt, K.: Gridchess: combining optimistic pondering with the young brothers wait concept. ICGA J. **35**(2), 67–79 (2012)
9. Himstedt, K., Lorenz, U., Möller, D.P.F.: A twofold distributed game-tree search approach using interconnected clusters. In: Luque, E., Margalef, T., Benítez, D. (eds.) Euro-Par 2008. LNCS, vol. 5168, pp. 587–598. Springer, Heidelberg (2008)
10. Hoki, K., Kaneko, T., Yokoyama, D., Obata, T., Yamashita, H., Tsuruoka, Y., Ito, T.: Distributed-shogi-system Akara 2010 and its demonstration. Int. J. Comput. Inf. Sci. **14**(2), 55–63 (2013)
11. Kaneko, T., Tanaka, T.: Distributed game tree search and improvements – match between hiroyuki miura and GPSShogi. IPSJ Mag. **54**(9), 914–922 (2013). (In Japanese)
12. Kishimoto, A.: Transposition table driven scheduling for two-player games. M.Sc. thesis, University of Alberta, January 2002
13. Knuth, D.E., Moore, R.W.: An analysis of alpha-beta pruning. Artif. Intell. **6**(4), 293–326 (1975)
14. Marsland, T.A., Popowich, F.: Parallel game-tree search. IEEE Trans. Pattern Anal. Mach. Intell. **7**, 442–452 (1985)
15. Obata, T., Sugiyama, T., Hoki, K., Ito, T.: Consultation algorithm for computer shogi: move decisions by majority. In: van den Herik, H.J., Iida, H., Plaat, A. (eds.) CG 2010. LNCS, vol. 6515, pp. 156–165. Springer, Heidelberg (2011)

16. Tanaka, T., Kaneko, T.: Massively parallel execution of shogi programs. In: The Special Interest Group Technical Reports of IPSJ. 2, vol. GI-24, pp. 1–8 (2010) (In Japanese)
17. Tsuruoka, Y., Yokoyama, D., Chikayama, T.: Game-tree search algorithm based on realization probability. ICGA J. **25**(3), 145–152 (2002)
18. Ura, A., Yokoyama, D., Chikayama, T.: Two-level task scheduling for parallel game tree search based on necessity. J. Inf. Process. **21**(1), 17–25 (2013)

Transfer Learning by Inductive Logic Programming

Yuichiro Sato[1][✉], Hiroyuki Iida[1], and H.J. van den Herik[2]

[1] School of Information Science, Japan Advanced Institute of Science and Technology, 1-1 Asahidai, Nomi, Ishikawa, Japan
{sato.yuichiro,iida}@jaist.ac.jp
[2] Leiden Institute of Advanced Computer Science, P.O. Box 9512, 2300 RA Leiden, The Netherlands
jaapvandenherik@gmail.com

Abstract. In this paper, we propose a Transfer Learning method by Inductive Logic Programing for games. We generate general knowledge from a game, and specify the knowledge so that it is applicable in another game. This is called Transfer Learning. We show the working of Transfer Learning by taking knowledge from Tic-tac-toe and transfer it to Connect4 and Connect5. For Connect4 the number of Heuristic functions we developed is 30; for Connect5 it is 20.

1 Introduction

An important property of a learning process is generalization. Consequently, a learning process is seen as an intelligent system able to generalize knowledge. For example, if a person has learned to play a game well, then that person is able to transfer his[1] knowledge to similar games. This means that the person is able to learn general knowledge about one game and then apply it as specific knowledge in another game. Based on this observation, we formulate the following research question: How do we construct a game-playing AI that has the same ability to adapt to new games as a human being? In other words, how can we transfer knowledge which is learned from one game to another game?

For this purpose, General Game Playing (GGP) is an appropriate research topic. A general Game Player is able to play, in principle, all discrete, finite and perfect information games (defined by General Games) without any human intervention [1]. There exist many successful implementations for a General Game Player [2–4]. GGP is a good test bed of algorithms generating game knowledge automatically. An example of such a generation for alpha-beta search and UCT-MC is reported by Walędzik and Mańdziuk [5]. Moreover, game knowledge generation for Heuristic functions that are produced by neural networks is described by Michulke and Thielscher [6]. However, these studies do generate game knowledge for a specific game with the aim to play that game well. What we are trying to achieve is (1) learning general knowledge from a game, and then

---

[1] For brevity, we use 'he' and 'his', whenever 'he or she' and 'his or her' are meant.

© Springer International Publishing Switzerland 2015
A. Plaat et al. (Eds.): ACG 2015, LNCS 9525, pp. 223–234, 2015.
DOI: 10.1007/978-3-319-27992-3_20

(2) applying the acquired knowledge as specific knowledge to another game. This is called Transfer Learning. Transfer Learning is a learning strategy that transfers previously learned general knowledge to improve the learning speed of a new game [7]. In GGP, Hinrichs and Forbus have reported Transfer Learning *by analogy* [8].

A telling example is learning the power of an additional square. This knowledge can be transferred to another domain. For instance, consider the domain of Tic-tac-toe. The game theoretical value of this game is draw. An interesting question is: What is the game theoretical value when we add an additional square as shown in Fig. 1? When using this board, the game is a win for the first player, (start at square 9, with the threat to play on square 8; the idea is to use the diagonal 2-6-10 as additional threat).

After this learning example, we consider the game of Chess. It is well-known that a king and two knights are unable to force mate. The highest goal to reach is stalemate. Assume that we augment the chess-board by an additional square e0. Then the question again reads: What is the game theoretical outcome of the KNNK endgame on this board? The answer reads: It is a win for the KNN side. The end position is shown in Fig. 2. The important point for transfer learning is that the power of an additional square in one game (Tic-tac-toe) may also unexpectedly change the original game theoretical outcome in another game (Chess). See also [8]. We invite readers to find analogous transfer ideas of this kind.

In this paper we apply the Inductive Logic Programming (ILP) approach to learn general knowledge for General Games. ILP is a successful approach, e.g., learning Chess variants and rules is reported to be possible [9]. Some ILP algorithms are able to make a reasonable specialization from general knowledge by winning examples only. If the examples represent normal winning situations, a winning strategy is expected to be learned. In our method, the general knowledge are boolean functions which represent patterns in a game position. The patterns may be winning position and losing position. The generated general patterns are then made specific for incorporation in Heuristic functions that apply to another game. This is an example of Transfer Learning between games.

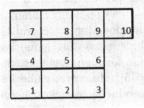

Fig. 1. A Tic-tac-toe game board with an additional square

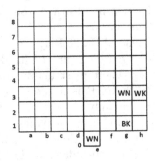

Fig. 2. A Chess game board with an additional square

In this paper, Tic-tac-toe is chosen as the source game, Connect4 and Connect5 are chosen as target games. We attempt to transfer general knowledge that is learned from Tic-tac-toe to Connect4 and Connect5. In Sect. 2, we define the general source concepts in such a way that they are suitable for transfer. In Sect. 3, we generate the concepts that will be transferred from Tic-tac-toe. In Sect. 4, we explain how ILP and Transfer Learning work. In Sect. 5, we transfer concepts that are learned from Tic-tac-toe to Connect4 and Connect5 in order to generate Heuristic functions for the game involved. In Sect. 6, we test the performance of the generated Heuristic functions. Section 7 provides a discussion. Section 8 concludes the paper.

2   Concepts in General Games

In GGP, games are described by a specific language, the Game Description Language (GDL). GDL is a Lisp-like language which has sufficient keywords to define General Games. General Games are discrete and finite; therefore, a game position is described as a finite set of pieces which have finite arguments. They form a string. If the game needs natural numbers, for example the x and y coordinates of a piece, they are defined in a succ(essor) relationship by the language [2].

There exist many types of concepts in General Games. For example, a pattern in a game position must be a sort of concept in that game. All patterns have a meaning. Some patterns indicate a close-to-win situation, while other patterns have a meaning as close to loss. This must be included in the set of all concepts of a particular game.

In GGP, patterns in General Games are also described by GDL. We are able to convert GDL to Prolog. Therefore, pattern matching of logic programming is able to describe patterns in General Games. Let us denote a piece in a game position as a *proposition* (with the name *piece*). For example, in Tic-tac-toe, all pieces are characterized by four coordinates (see Fig. 3). The first coordinate is the type, in our case it is a cell. The second and third coordinates are the x and y coordinates. The fourth coordinate is the occupation (x, o, or blank).

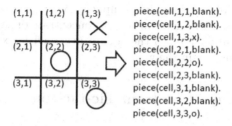

Fig. 3. A Tic-tac-toe game position as propositions

In Fig. 3, the first two pieces are described as

```
piece(cell,1,1,blank).
piece(cell,1,2,blank).
```

These two pieces are adjacent since the x coordinate is the same and the y coordinate differs by one. Let us introduce four variables (viz. C, X, Y, and S) to generalize a pattern in this position. Next to the above propositions, we also have arithmetic propositions such as $Y2$ is $Y + 1$ (notation is in Prolog). Now consider the following pattern.

```
patternX :- piece(C,X,Y,S), Y2 is Y + 1, piece(C,X,Y2,S).
```

This pattern is a conjunction of three propositions (a conjunction is a combination of propositions connected by and; in the example represented by a comma). It is applicable to any game which has a two-dimensional game board and symbols on it. If there exist two adjacent pieces which are represented by the same symbol, this pattern returns true. In Fig. 4, we find in 4a and 4b the pattern of two adjacent oo in the top row. In Fig. 4c, we see this pattern in the second row (seen from the bottom) on position four and five. This way of characterizing patterns is useful to distinguish game positions and is part of the semantics of all games.

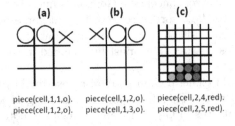

Fig. 4. A concept on Tic-tac-toe and Connect4

In games, there exist also other types of patterns. Examples are (1) a disjunction of propositions, and (2) patterns of time evolution of positions, e.g., a sequence of changes of positions. For simplicity, we focus only on non-complex patterns in a position. From now on, let us concentrate on straightforward patterns in a game position (i.e., is the square occupied by x, o, or blank) and consider them as concepts in the games.

3   Concept Generation from Tic-tac-toe

It is possible to generate concepts from game simulations. In this section, we generate concepts from simulations of Tic-tac-toe. These concepts are useful to play other games, e.g., Connect4 and Connect5. Below, we investigate the generation of conjunctions and disjunctions.

Our procedure is as follows. We generate conjunctions by replacing the same symbols in the arguments by a variable. If the n-th arguments of pieces are the same, then they are replaced by a variable. If the n-th arguments of pieces are a number, a and b, then a is replaced by a variable and b is replaced by the sum of the variable and $b - a$ as is done in previous work [5]. The total result of concept generation from random game simulations of Tic-tac-toe is seen in Fig. 5. We generated two types of concepts. One is a binary concept and the other one is a ternary concept. Binary concepts are patterns with two pieces. Ternary concepts are patterns with three pieces. Concepts are learned from positions after a playout. We see that if the number of simulations increases the number of generated concepts also increases. After 2,000 simulations, the learning process is saturated, i.e., 81 concepts are generated.

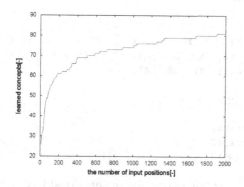

Fig. 5. Conjunction concept generation from Tic-tac-toe simulation

A disjunction of propositions is generated by Algorithm 1 (a disjunction is a combination of propositions connected by or). The algorithm has two types of parameters, viz. concepts and positions. The input concepts are conjunctions that are generated from Tic-tac-toe simulations as above. The input positions are random simulations of Tic-tac-toe games with a winner. The algorithm generates a disjunction which matches the input positions maximally, and has a quadratic running time. The result is given in Fig. 6. As the number of input positions grows, the number of conjunction concepts in a disjunction also grows. After 2,000 simulations, the learning process is saturated. Finally, a disjunction made of 17 conjunctions is generated. Three examples in which conjunctions are involved are as follows.

```
concept1 :- piece(X1, X2, X3, X4), X5 is X3 - 1, piece(X1, X2, X5, X4),
            X6 is X5 - 1, piece(X1, X2, X6, X4).
concept2 :- piece(X1, X2, X3, X4), X5 is X3 + 1, piece(X1, X5, X3, X4),
concept3 :- piece(X1, X2, X3, X4), X5 is X3 - 2, X6 is X5 + 1,
            piece(X1, X6, X3, X4).
```

One of the most complex learned pattern, learned from Tic-tac-toe, is as follows.

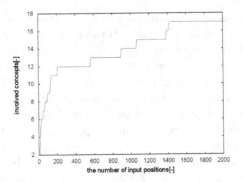

Fig. 6. Disjunction concept generation from Tic-tac-toe simulation

```
concept11:- piece(X1, X2, X3, X4), X5 is X2 + 1, X6 is X3 + 1,
            piece(X1, X5, X6, X4), X7 is X2 + 2, X8 is X3 + 2,
            piece(X1, X7, X8, X4).
```

The example disjunction reads

```
disjunction :- concept1.
disjunction :- concept2.
disjunction :- concept3.
```

The concepts (i.e., conjunctions and disjunctions) that are generated in this section are used to make a Heuristic function by ILP.

4   Concept Specialization by ILP

ILP is a research topic that generates theorems automatically [10]. For example, it may create a theorem from positive examples, negative examples, and background knowledge. In our case, it is a specialization in the target game that is built up by a number of general concepts from the source game. Concepts that are generated in the previous section are too general to play a specific role in that target game. Therefore they should be specified. For example, assume the following concept (concept Y) is learned by generalization of a final position in Tic-tac-toe (the position is assumed to be a win or a loss). For instance, take concept Y as follows.

```
conceptY :- piece(X1, X2, X3, X4), X5 is X3 + 1, piece(X1, X2, X5, X4),
            X6 is X5 + 1, piece(X1, X2, X6, X4).
```

Let us denote the following set of pieces as position 1.

```
piece(cell, 1, 1, o).
piece(cell, 1, 2, o).
piece(cell, 1, 3, o).
```

Algorithm 1. disjunctionGeneration(*concepts*, *positions*)

---

restPosition $\Leftarrow$ *positions*
restConcepts $\Leftarrow$ *concepts*
result $\Leftarrow$ empty list
while 0 < size of *restPositions* or 0 < size of *restConcepts* **do**
   counts $\Leftarrow$ count matchings for each *restConcepts* to *restPositions*
   if 0 < max element of *counts* **then**
      maxConcept $\Leftarrow$ choose a concept with maximum matching from *restConcepts*

      append *maxConcept* to *result*
      remove *maxConcept* from *restConcepts*
      remove positions which matches to *maxConcept* from *restPositions*
   else
      return make disjunction of *result*
   end if
end while
return make disjunction of *result*

Subsequently, let us denote the following set of pieces as position 2.

```
piece(cell, 1, 1, x).
piece(cell, 1, 2, x).
piece(cell, 1, 3, x).
```

The concept Y is true for both position 1 and position 2.

Assume there exist two players: the o player and the x player. From concept Y, it is impossible to evaluate whether o's line is good and x's line is bad; both lines are evaluated as the same. To distinguish the o line from the x line, the concept should be specified. In this case, $X4$ should be replaced by o or x. ILP is useful to make (1) this kind of specialization and (2) to formulate Heuristic functions for the specialization made under (1).

ILP algorithms find the most fitting proposition that explains the examples by background knowledge. In this case, assume position 1 is a positive example and position 2 is a negative example. Moreover, concept Y is taken as background knowledge. Then, the ILP algorithm finds the only difference between position 1 and position 2, being the fourth argument. Consequently, it will make the following specialization (i.e., the winning specialization is: replacing $X4$ by o).

```
conceptY :- piece(cell, 1, 1, o), X5 is 1 + 1, piece(cell, 1, X5, o),
            X6 is X5 + 1, piece(cell, 1, X6, o).
```

This specialization satisfies our demand.

In the literatures we found many ILP implementations. For our ILP tool, we used Aleph [11]. Aleph is a tool described by Muggleton and De Raedt [12,13]. What Aleph does is specifying general concepts that are learned from Tic-tac-toe simulations in relation to a target game.

We tried two specializations, viz. a specialization to Connect4 and to Connect5. Positive examples are game positions that end in a win; negative examples are game positions that end in a loss. In these process, general knowledge is transferred from a simple game (Tic-tac-toe) to more complicated games (Connect4 and Connect5).

5   Transfer Learning by Concept Specialization

We tried Transfer Learning by specializing concepts from Tic-tac-toe to Connect4 and Connect5. Positive and negative examples are given by random game simulations. Specialized concepts are (1) a set of conjunctions and (2) a disjunction made of the conjunctions (see Sect. 3).

For Connect4, we experimented with a different number of positive examples (and similarly negative examples). The range of the number of positive and negative examples ran from 1 to 20 for conjunctions; and from 1 to 10 for the best disjunction (we used only one disjunction in our experiments).

For Connect5, specializations were performed only for conjunctions. The range of the number of positive and negative examples ran again from 1 to 20. For each set of positive and negative examples, a different specialization was obtained.

In both cases, we see that if the number of examples increases, the number of generated specified concepts also increases (see Fig. 7).

Let us now provide an example of a specified concept that is obtained by the above ILP process. A Heuristic function generated by specialization of conjunctions by 20 positive and 20 negative examples for Connect4 is made of the following 5 specified conjunctions (see Fig. 8), consisting of concept 4, 6, 8, 10, and 11.

```
concept4 :- piece(cell, 3, 2, r), Y1 is 3 + -1, Y2 is 2 + -1,
                piece(cell, Y1, Y2, r).
concept6 :- piece(cell, 5, 3, w), Y1 is 5 + 2, Y2 is 3 + -1,
                piece(cell, Y1, Y2, w).
concept8 :- piece(cell, 3, 2, b), Y1 is 3 + -1, piece(cell, Y1, 2, b).
concept10 :- piece(cell, 1, 2, r), Y1 is 1 + 2, Y2 is 2 + 2,
                piece(cell, Y1, Y2, r).
concept11 :- piece(cell, 1, 3, w),  Y1 is 1 + 2, Y2 is 3 + 1,
                piece(cell, Y1, Y2, w).
```

The Heuristic function is made for the first player when playing Connect4. Let us have a closer look at the specifications. We take concept 11. This concept obtained the specification that the type of piece is characterized by cell; the x coordinate is specified by 1, the y coordinate is specified by 3 and the occupation by w that is the symbol for the white player (first player) in Connect4 (second player is r).

For specialization toward Connect4, all conjunctions and the disjunction that are obtained by Tic-tac-toe were used as background knowledge. However, not

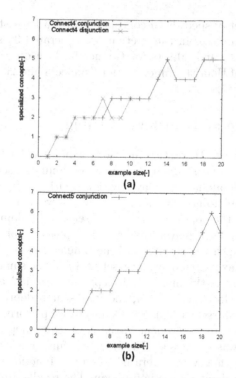

Fig. 7. Specialization to Connect4 and Connect5

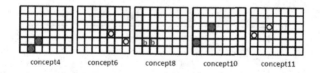

Fig. 8. A Heuristic function for Connect4 seen as a set of specified concepts

all of them were used for the specialization. This means that some concepts are useful, but others are not useful for this type of game. Here, we may anticipate on the difference in complexity of Connect4 and Connect5. For instance, we may state that, in a Tic-tac-toe specialization process toward Connect5, only concepts which appear in the specialization for Connect4 are used as background knowledge. This is a meta-concept, i.e., a relationship between concepts. The meta-concept is suitable for reducing the computation time.

Once general concepts are specified to a target game, the specialized concepts are useful to make a Heuristic function for that game. Our Heuristic functions are a set of specialized concepts. If a specialized concept is true in a position, we may evaluate the position; it will have a positive constant value. If some specialized concepts are true in a position, the evaluated value of the position is the total sum of the constant values.

In summary, a set of specialized concepts creates a Heuristic function. From the specialization processes in this section, we generated 30 sets of specialized concepts for Connect4 and 20 sets for Connect5. As a direct consequence, we generated 30 types of Heuristic functions for Connect4 and 20 types of Heuristic functions for Connect5.

6   Performance of Transfer Learning

The performance of the Heuristics functions generated by the specializations for Connect4 and Connect5 were tested by game simulations. Game simulations were performed by an alpha-beta player with a Heuristic function against a random player (or an alpha-beta player). For Connect4 we had four experiments. In experiment 1 and 2, the opponent player was set as a random player. In experiment 3 and 4, the opponent player was set as an alpha-beta player. In experiment 1, the max search depth was set at 1 and the simulation size was set at 5,000. In experiment 2, the max search depth was set at 3 and the simulation size was set at 500. In experiment 3, the max search depth was set at 1 and the simulation size was set at 500. In experiment 4, the max search depth was set at 3 and the simulation size was set at 500. For Connect5 we performed only one experiment (experiment 5). In experiment 5, the max search depth was set at 1 and the simulation size was set at 500. The Heuristic functions were indexed by the number of examples that were generated. Heuristic function 0 means the use of alpha-beta search without Heuristic function. The results are seen in Figs. 9, 10 and 11.

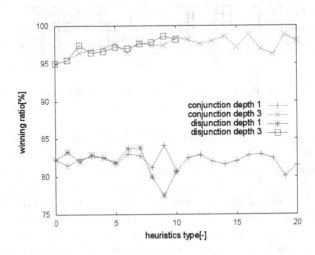

Fig. 9. Game simulation with a player using a Heuristic function vs a random player for Connect4

There exists a tendency that the winning ratio increases as the index of the Heuristic functions also increases (see Figs. 9, 10 and 11). This means that the

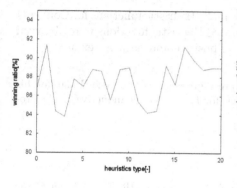 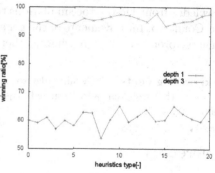

Fig. 10. Game simulation with heuristics player vs random player for Connect5, search depth 1

Fig. 11. Game simulation with heuristics player vs 1-depth alpha-beta player for Connect4

Heuristic functions that were generated by many positive and negative examples, have a better performance when compared to Heuristic functions that were generated by a smaller number of positive and negative examples. The tendency clearly appears for the depth-3 search case. We note that depth-1 searches totally rely on Heuristic functions. However, our Heuristic functions are not perfect, they have inaccuracies. Therefore we surmise that they guide the middle games successfully, but miss sometimes a win in the endgame. This is why our Heuristic functions perform better in 3-depth search than in 1-depth search, even though the same Heuristic function is used.

7   Discussion

We observed Transfer Learning between Tic-tac-toe, Connect4, and Connect5. The success relies on the fact that these games have a similar structure. For example, the games have a two-dimensional game board, the goal of the game is to make a line on the board and once a player puts a mark on the board, the mark never moves.

To do more general Transfer Learning, more analysis for the semantics of games is needed, e.g., the role of pieces needs to be analyzed. In other games, there exist many pieces with a specific role. For example, in Chess and Shogi (Japanese Chess) some pieces have the same legal moves but others do not. If the similarity between pieces with the same movements has been learned, more general Transfer Learning will be available.

8   Conclusions

In this paper, we successfully observed Transfer Learning by ILP between games that have a similar structure. It was possible to produce a general background knowledge from Tic-tac-toe simulations to make Heuristic functions for Connect4 and Connect5. Improvements of the generated Heuristic functions were observed when we prepared an increasing number of positive and negative examples.

Acknowledgments. We would like to express great thanks to Aske Plaat for his advice to this research, and Siegfried Nijssen for his advice on Inductive Logic Programming.

References

1. Love, N., Hinrichs, T., Haley, D., Schkufza, E., Genesereth, M.: General Game Playing: Game Description Language Specfication. Technical report LG-2006-01 Stanford Logic Group (2006)
2. Schiffel, S., Thielscher, M.: Fluxplayer: a successful general game player. In: The Twenty-Second AAAI Conference on Artificial Intelligence, pp. 1191–1196 (2007)
3. Björnsson, Y., Finnsson, H.: CADIAPLAYER: a simulation-based general game player. IEEE Trans. Comput. Intell. AI Games **1**(1), 4–15 (2009)
4. Méhat, J.M., Cazenave, T.: Ary, a general game playing program. Board Games Studies Colloquium (2010)
5. Walędzik, K., Mańdziuk, J.: An automatically-generated evaluation function in general game playing. IEEE Trans. Comput. Intell. AI Games **6**(3), 258–270 (2014)
6. Michulke, D., Thielscher, M.: Neural networks for state evaluation in general game playing. In: Buntine, W., Grobelnik, M., Mladenić, D., Shawe-Taylor, J. (eds.) ECML PKDD 2009, Part II. LNCS, vol. 5782, pp. 95–110. Springer, Heidelberg (2009)
7. Taylor, E.M., Stone, P.: Transfer learning for reinforcement learning domains: a survey. J. Mach. Learn. Res. **10**, 1633–1685 (2009)
8. Hinrichs, R.T., Forbus, D.K.: Transfer learning through analogy in games. AI Mag. **32**(1), 70–83 (2011)
9. Muggleton, S., Paes, A., Santos Costa, V., Zaverucha, G.: Chess revision: acquiring the rules of chess variants through fol theory revision from examples. In: De Raedt, L. (ed.) ILP 2009. LNCS, vol. 5989, pp. 123–130. Springer, Heidelberg (2010)
10. Mitchell, M.T., Keller, M.R., Kedar-cabelli, T.S.: Explanation-based generalization: a unifying view. Mach. Learn. **1**(1), 47–80 (1986)
11. Aleph. http://www.cs.ox.ac.uk/activities/machlearn/Aleph/aleph.html
12. Muggleton, S.H., De Raedt, L.: Inductive logic programming: theory and methods. J. Logic Program. **19–20**, 629–679 (1994)
13. Muggleton, S.: Inverse entailment and progol. New Gener. Comput. **13**(3–4), 245–286 (1995)

Developing Computer Hex Using Global and Local Evaluation Based on Board Network Characteristics

Kei Takada[✉], Masaya Honjo, Hiroyuki Iizuka, and Masahito Yamamoto

Graduate School of Information Science and Technology, Hokkaido University,
Kita 14, Nishi 9, Kita-ku, Sapporo, Hokkaido 060-0814, Japan
{takada,honjyo,iizuka,masahito}@complex.ist.hokudai.ac.jp

Abstract. The game of Hex was invented in the 1940s, and many studies have proposed ideas that led to the development of a computer Hex. One of the main approaches developing computer Hex is using an evaluation function of the electric circuit model. However, such a function evaluates the board states only from one perspective. Consequently, it is recently defeated by the Monte Carlo Tree Search approaches. In this paper, we therefore propose a novel evaluation function that uses network characteristics to capture features of the board states from two perspectives. Our proposed evaluation function separately evaluates the board network and the shortest path network using betweenness centrality, and combines the results of these evaluations. Furthermore, our proposed method involves changing the ratio between global and local evaluations through a support vector machine (SVM). So, it yields an improved strategy for Hex. Our method is called Ezo. It was tested against the world-champion Hex program MoHex. The results showed that our method was superior to the 2011 version of MoHex on an 11×11 board.

1 Introduction

Hex is a classic board game. Classified as a two-player, zero-sum, perfect information game, it was independently invented by Piet Hein and John Nash [1]. Hex is played on a rhombic board consisting of hexagonal cells. An 11×11 board is traditionally used, but it can be any size (Fig. 1 shows a 7×7 board). The two players have uniformly colored pieces (e.g., black and white), and the game proceeds by players placing their pieces in turn on empty cells. The two black opposing sides of the board are assigned to the black player, and other two opposing sides are assigned to the white player. The goal of the game is connecting the two opposing sides by own color pieces: the black player wins if the black player successfully connects the black sides using black pieces, whereas the white player wins if the white player successfully connects the white sides using white pieces (Fig. 2). It is shown that the first player has a winning strategy [2], and the game cannot end in a draw [3].

© Springer International Publishing Switzerland 2015
A. Plaat et al. (Eds.): ACG 2015, LNCS 9525, pp. 235–246, 2015.
DOI: 10.1007/978-3-319-27992-3_21

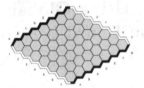

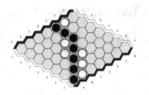

Fig. 1. 7 × 7 Hex gameboard

Fig. 2. Game board showing a winning configuration for black

Approaches to the development of computer Hex can be roughly classified into two groups: those that use evaluation functions (e.g., WOLVE [4], SIX [5], HEXY [6]), and those that use the Monte Carlo Tree Search (MCTS) (e.g., MOHEX [7], PANORAMEX [8]). A well-known program that uses an evaluation function is WOLVE developed by Arneson [4]. The evaluation functions used in WOLVE and similarly developed programs are based on the electric circuit model. This model involves evaluating board states only from one perspective. By evaluating board states from a greater number of perspectives, since human players change strategies in response to board states, we think that a better evaluation function can be created.

In this paper, we propose a novel evaluation function using two different perspectives, and develop a computer Hex program called EZO by dynamically changing the ratio between the strategies yielded by the two perspectives. Board states are regarded as a network in order to capture these from different perspectives. Our evaluation function separately evaluates board states from the board network and the shortest path network, and combines the two evaluations. Global evaluation involves evaluating the overall strategy of board states, and local evaluation evaluates local strategy directly relating to winning or losing the match. In strategizing for Hex, it may be effective to capture the global strategy at the beginning of the match, the local strategy becomes more important in the latter part of the match. It is possible to change the criteria of evaluation by changing the ratio between the global and local evaluations. This allows for the development of a satisfactory strategy for Hex. We verify the effectiveness of our novel evaluation function as well as the method to alter the ratio between the two evaluations by comparing its performance with that of MOHEX and WOLVE, respectively.

2   Proposed Method

We propose a novel evaluation function using network characteristics, and develop a computer Hex program based on the evaluation function [9]. We first describe the method of creating the board network from board states, followed by the evaluation function formulated by using network characteristics calculated from the board network. Finally, we detail our proposal and call it ComputerHex.

In this paper, we use players with black and white. Black is the first player and White is the second.

2.1   Board Network Including Extra Board State Information

The states of the Hex board can be expressed as a network by treating cells as nodes and connecting adjacent nodes with a link (or edge). The board network is first created in this manner for each player, and is used to view the global board states. The shortest path network between the sides of the board is then created for each player to view the local board states, which are related to the results of the matches. $G_B^b(V, E)$ is the board network and $G_B^p(V', E')$ is the shortest path network for Black, and $G_W^b(V, E)$ is the board network and $G_W^p(V', E')$ is the shortest path network for White, where V, V' are sets of nodes and E, E' are sets of links. In order to consider future board states several turns ahead, we use the idea of a *virtual connection* and *virtual semi-connection* [13]. Those connections are included in the set of links E or E'.

Virtual Connection. *Virtual connection* (*VC*) and *virtual semi-connection* (VSC) are connection strategies between two cell groups for the second player and the first player, respectively. A connection strategy connects two cell groups. *VSC* is a strategy whereby connections can be formed by placing the relevant piece appropriately in a given move. *VC* is a strategy whereby pieces can be connected even after an opponent's move. If there are *VCs* between the two sides of the board belonging to a player, it means that the player has a winning strategy. A large number of *VCs* are found by applying two deduction rules, called the *and-rule* and the *or-rule*. The *and-rule* can combine *VCs* to a create new *VCs* or *VSCs*, and the *or-rule* combines non-interfering *VSCs* to create new *VCs* (Fig. 3). The algorithm for finding new *VCs* or *VSCs* is called *h-search* [13]. We use *h-search* and pattern matching to create *VCs*. We augment our function by adding a few *VC* patterns that cannot be found by *h-search*, as published by King [14].

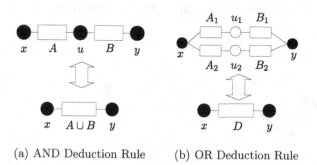

(a) AND Deduction Rule   (b) OR Deduction Rule

Fig. 3. Two deduction rules. x, y, u is a cell or cell group, and A, B, D is a node set that constitutes VC or VSC

2.2   Board Network Creation Algorithm

In order to evaluate the board state from different perspectives, we use two kinds of networks: the board network and the shortest path network. We define cell i as node v_i, and connect it to adjacent nodes with a link (or edge). V is a set of nodes and E is a set of links. Function C is defined as the condition of nodes. $C(v_i) = 0$ if v_i is an empty cell, $C(v_i) = 1$ if v_i is occupied by Black, $C(v_i) = -1$ if occupied by White. Further, the two opposing sides belonging to each player are represented by nodes v_s and v_t. $C(v_s) = 1$ and $C(v_t) = 1$ for sides belonging to Black, and $C(v_s) = -1$ and $C(v_t) = -1$ for those belonging to White.

The process to create a board network $G_B^b(V, E)$ for Black is shown below. The board network for White $G_W^b(V, E)$ can be obtained in an analogous manner, replacing black by white.

1. Links $e(v_i, v_j)$ are added to E between all nodes adjacent to v_i and v_j.
2. Links $e(v_i, v_s)$ are added to E between all nodes v_i adjacent to v_s, and $e(v_j, v_t)$ is added to E between all nodes v_j adjacent to v_t.
3. Nodes belonging to White v_i ($C(v_i) = -1$) are removed from V, and links $e(v_i, v_j)$ belonging to the v_i are removed from E.
4. The *h-search* algorithm and pattern matching are applied to the board network, and the VCs yielded are added to E.

Figure 4 shows an example of a board network. The shortest path networks $G_B^p(V', E')$ and $G_W^p(V', E')$ are created by nodes and links that form the shortest path between v_s and v_t in order to evaluate the local strategy directly relating to winning or losing the match.

2.3   Evaluation Function Using Network Characteristics

We propose an evaluation function based on network characteristics calculated using the board networks and the shortest path networks that we created. The evaluation function consists of global and local evaluations.

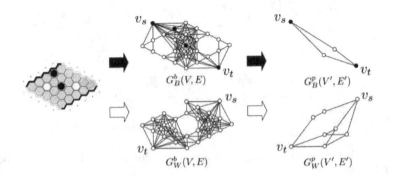

Fig. 4. Example of a 5×5 board state (left). The middle diagram shows the board network $G^b(V, E)$. The right diagram shows the shortest path network $G^p(V', E')$ between v_s and v_t.

The objective of Hex to create a path between opposing sides, thus considering the variety of paths between any two nodes is important for successful Hex strategies. To evaluate the global strategy on the board, we use the average of betweenness centralities in the board network, which is an index of how a node contributes to shortest paths in the network. Betweenness centrality can express the variety of possible paths. On the contrary, the maximum value of betweenness centrality in the shortest path network is used to evaluate local strategy directly related to winning or losing the match. The net evaluation function E_v is defined as follows:

$$Ev = (1 - \alpha)\frac{C_W}{C_B} + \alpha\frac{C'_W}{C'_B},\tag{1}$$

where C_B is the average of betweenness centralities of Black's board network $G^b_B(V, E)$, and C_W represents this for White's network. C'_B is the maximum value of betweenness centrality in the shortest path network $G^p_B(V', E')$ between v_s and v_t for Black, and C'_W is that for White. α is a constant parameter used to adjust the weight of the global and local strategies.

Betweenness centrality of node v_i is calculated as follows,

$$b_{v_i} = \frac{2}{(N - 1)(N - 2)}\sum_{k=1}^{N}\sum_{j=1}^{N} g(v_i, v_k, v_j),\tag{2}$$

where N is the number of nodes in the board network, and $g(v_i, v_k, v_j)$ is a function that returns 1 when node v_i is included in the shortest path network between v_k and v_j, and 0 otherwise. In case $v_i = v_k$, $v_j = v_k$, or $v_i = v_j$, $g(v_i, v_j, v_k)$ is 0.

Boarding states with small values of C_B or C_W imply the existence of many paths between any pair of nodes and many global strategies on the board. Thus, small values of C_B and large ones of C_W indicate favorable states for Black from the global perspective, and unfavorable ones for White. Figure 5 shows an example of board states with small values of C_B and large ones of C_W. Moreover, the board states for small values of C'_B or C'_W imply the existence of a large number of paths as candidates for the shortest path between v_s and v_t, and many

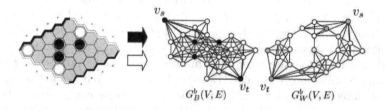

Fig. 5. Example of a 5×5 board state (left).The board network $G^b_B(V, E)$ has a small value C_B, and $G^b_W(V, E)$ has a large C_W. It means that Black is playing better than White.

local strategies relating to the result of the match. In an analogous manner, small values of C_B' and large ones of C_W' imply states favorable to Black with regard to the local perspective related to the result. In sum, the global evaluation is performed by the first term, C_W/C_B, and the local evaluation by the second term, C_W'/C_B'. The constant parameter α adjusts the contribution weight of the global and local evaluations.

2.4   Computer Hex Using E_v

We propose computer Hex CH_{E_v} based on the above evaluation function. CH_{E_v} uses Ev as the evaluation function and a 2-ply $\alpha\beta$ search algorithm as a game-tree search algorithm. In the electric circuit model, it is known that nodes with higher energies are assigned higher priorities with regard to placing pieces. Similarly, nodes with the high values of betweenness centrality are easy to place pieces on. Hence, we use move ordering based on betweenness centrality. We adopted 2-ply search because considerably more time is required to search for the best move using 3-ply or 4-ply search. The search time for our proposed method using 2-ply was approximately 35 s. In CH_{E_v}, α is a constant value.

3 CH_{E_v} with Fixed α vs. MoHex

In order to evaluate our proposed method CH_{E_v} for Hex, we tested it against MoHex. MoHex is an MCTS player and the reigning Computer Olympiad Hex gold medalist [11]. We downloaded the 2011 version of MoHex from the Benezene project website[1]. MoHex has restricted the time to search future moves to within 10 s.

CH_{E_v} played against MoHex on an 11×11 board 100 times for each value of α and each player (the first and the second player). The swap rule was not considered. Each player started the search from the opening move. The game judgment was performed by a 10-second Depth-First-Proof-Number (DFPN) search downloaded from the website after every move [12]. The computer used had a Phenom II X6 processor (six cores, 2.9 GHz clock).

Figure 6 shows the winning percentages of our method against MoHex for each value of α. The highest winning percentage was obtained when $\alpha = 0.075$ (79 %) for the first player and $\alpha = 0.05$ (18 %) for the second player.

The highest winning percentage ($\alpha = 0.075$) for our method was considerably higher than that for a method that only used a global evaluation function ($\alpha = 0$). This means that both the global and local evaluations of our proposed method worked in an appropriate manner. This was also true for the second player, although the difference was small. Despite the fact that our evaluation function only calculated simple network characteristics, e.g., betweenness centrality, etc., the winning percentage of the first player against MoHex was surprisingly high.

---

[1] http://benzene.sourceforge.net/.

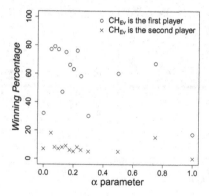

Fig. 6. CH_{E_v} with fixed α vs. MoHex (100 trials for each parameter).

4 Improving the Proposed Method

We have shown that our proposed evaluation function, which combines global and local evaluations, is effective. However, the winning percentage of CH_{E_v} against MoHex was 79 % of the first player and 18 % for the second player. In other words, the winning percentage of MoHex against CH_{E_v} was 82 % for the first player and 21 % for the second player. Consequently, CH_{E_v} is weaker than MoHex.

We propose an extended evaluation method that alters the value of α in response to board states. The local evaluation is important in determining the winner of a match. We expect that it would be more effective to increase the influence of the local evaluation in the latter part of matches because it is possible for a match to last only a few turns. The value of α determines the ratio between the global and local evaluations, and it is possible to increase the influence of the local evaluation by increasing the value of α. In this section, we describe a method to detect the time at which the value of α is changed using network characteristics (4.1). Subsequently, we investigate the timing of the changes by a support vector machine (4.2).

4.1 Effectiveness of Increasing α

In order to show that a higher value of α is effective in the latter part of matches, we compare the winning percentages of the evaluation function with high and low values of α. A high value of α was set to 0.5, and low values of α were set to 0.075 for the first player and 0.05 for the second player. These last two values were the best ones obtained in Sect. 3. The initial board states were given by the history of matches between CH_{E_v}, with fixed low values of α, and MoHex. Therefore, we could compare the winning percentages of the evaluation functions with fixed high and low values of α for board states during matches. The comparison of winning percentages clarified, for a given board state, whether a low or high value of α should subsequently have been used. There were 659 board states for the

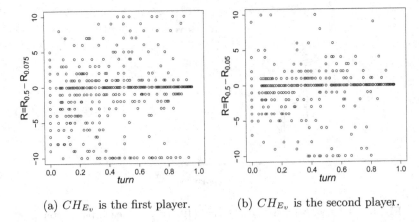

(a) CH_{E_v} is the first player.   (b) CH_{E_v} is the second player.

Fig. 7. The difference in the number of wins between large α and small α.

first player and 655 for the second player in the history of the matches. Starting from the board states, we compared the winning percentages. To calculate these, we tested the matches starting from the same board states 10 times.

Figure 7 shows the difference between the number of wins obtained by using keep-low α and those obtained using change-to-high α. The horizontal axis shows the normalized number of turns, where 1.0 represents the end of a match. The vertical axis shows the difference in the number of wins between high values of α ($\alpha = 0.5$) and low values ($\alpha = 0.075$ or $\alpha = 0.05$). This means that the plots of high values at the vertical axis show that the winning percentages increased by changing to higher α following the relevant board states. In the extreme case, where the difference was 10, only high values of α could win against MoHex and low values could never win following the relevant board states. The results showed that there were some cases where α should have been changed to higher values in the latter parts of matches.

It was clear that changing α can be effective. However, the kinds of board states that should be changed and the appropriate time to increase the value of α remain vague. The results also showed that changing the value of α at inappropriate times can be detrimental. The next section contains an explanation of a method to detect the timing of changes in values of α.

4.2   Creating Classifier by Using SVM

In order to capture features of the board states, the following 12 network characteristics (six for each player) were used:

- the maximum, minimum, variance, and average of betweenness centralities over all nodes of the board network (i.e. four characteristics),
- the maximum values of betweenness centrality in the shortest path network (C'_B, C'_W);
- the shortest path length between v_s and v_t (d_B, d_W).

Each betweenness centrality (max, min, variance) measures the biases of the relevant strategy. The average of betweenness centrality (C_B, C_W) captures the global strategy, and C'_B and C'_W capture the local strategy. d_B and d_W estimate how close a player is to victory.

Using the 12 network characteristics described above, we developed two classifiers for the first and the second player in order to classify board states into cases where the values of α should or should not be changed. For this purpose, we used a support vector machine (SVM), which is well-known as a high-precision binary classifier and a method for supervised machine learning [15]. It learns labels using training data and estimates these labels for new data.

The training data were collected from the results of Sect. 4.1. Higher values on the vertical axis imply that α should have been changed to high values, and low values mean that α should not have been changed in the relevant board states. Therefore, we assigned a positive label (changing α) to a board state if the value in Fig. 7 was greater than 4, and negative labels (maintaining the value of α) to the other board states. The number of the board states satisfying $R > 4$ was 42 for the first player and 12 for the second player. SVM learning used the "kernlab" library in the statistical analysis software R, and performed a non-linear classification using a kernel function [16]. Each parameter for SVM was determined by a grid search.

5 Experiment

In order to show that our proposed computer Hex EZO, which involves dynamically changing the value of α, is superior to the fixed-α computer Hex, we compared EZO with WOLVE, which uses the best available evaluation function based on an electrical circuit model.

5.1 Proposed Method

EZO uses E_v as the evaluation function, a 2-ply $\alpha\beta$ search, and a classifier developed using an SVM, as described in Sect. 4.2. EZO uses move ordering based on betweenness centrality, and takes approximately 35 s to search for a next move. The value of α began at 0.075 for the first player or 0.05 for the second player. When the classifiers output $\alpha = 0.5$ from the network characteristics of the board states, the value of α was changed to 0.5. Following this, α was kept constant ($\alpha = 0.5$).

5.2 Conditions

The winning percentage of EZO against MOHEX was compared with the winning percentages of our fixed-α method against WOLVE. EZO was compared indirectly with WOLVE (through MOHEX) in this manner because the matches between the methods using evaluation functions were deterministic, and always generated the same moves. Hence, direct comparison was difficult. Because the algorithm

Table 1. The winning percentages of each computer Hex program for 100 trials (first/second player). ± represents standard error, 68 % confidence.

	Win % vs. MoHEX
EZO	92 ± 2.7/24 ± 4.2
CH_{E_v} (fixed $\alpha = 0.075$ or $\alpha = 0.05$)	79 ± 4.1/18 ± 3.8
WOLVE	82 ± 3.8/42 ± 4.9
MoHEX	83 ± 3.7/17 ± 3.7

of MoHEX was based on MCTS, the winning percentages against MoHEX could be compared.

The experiment used four computer Hex programs, i.e., EZO, the CH_{E_v} fixed-α method, WOLVE, and MoHEX. The 2011 version of WOLVE was downloaded from the Benezene project site[2]. The swap rule was not considered. Each player started the search from the opening move. The search time of WOLVE was within 10 s, as for MoHEX. The board size was 11 × 11, and 10-second DFPN search was used for game judgment after each move.

5.3   Result

Table 1 shows the winning percentages for each computer Hex player against MoHEX. The results show that the winning percentages increased by changing the value of α based on SVM for both the first and the second player. This means that SVM could properly recognize board states and decide when α changed. The number of matches where α changed was 19 for the first player and nine for the second player. EZO won 16 matches of 19, and eight out of nine for the first and the second player, respectively. The winning percentage of fixed-α was worse than that of WOLVE. However, our improved method won more than Wolve for the first player. Compared to MoHEX, EZO obtained a higher winning percentage.

5.4   Discussion

The results in Sect. 5 show that increasing the value of α, which represents a shift to the strategy, at a certain timing during the match is effective to obtain a higher winning percentage. It is a popular method that is often used in the other games, such as Go and Shogi (Japanese chess) [17]. For example, the phases of the game in Shogi are roughly classified into the opening game, the middle game, and the endgame. The transitions of the phases can be clarified and described to some extent by the rules in the long history of the studies. This is not yet clear in Hex. WOLVE and MoHEX do not explicitly use phase transition, though it is possible for the evaluation function to use it indirectly. In our method, the

---

[2] http://benzene.sourceforge.net/.

transition timing data was collected from repeated matches starting from the same board states, and the proper timings were trained using the SVM. Our results showed that a trained SVM can detect the timing of the transition even for unknown board states. Our proposed method has this advantage, and thus was able to attain a higher winning percentage against MoHex.

The winning percentage of our proposed method for the first player was better than that of Wolve, but was worse for the second player. According to their original paper [4], the creators of Wolve used the same evaluation function for the first and the second player. We also used the same evaluation function, although a different value of the parameter was chosen for each player. Despite the fact that our method was adjusted for the second player, the winning percentage of the second player could not overtake that of Wolve. There are two reasons for this.

One is that the evaluation function of Wolve was adjusted to win for both the first and the second player. It is known that the strategies of the first player are quite different from those of the second player because there always exist winning strategies at the beginning of the match for the first player but not for the second player. Unless the first player makes mistakes, i.e., loses the winning strategies, the first player can win. The second player must make moves to induce mistakes from the first player. Although the strategies are rather different, the evaluation function of Wolve might be able to evaluate moves for both strategies. On the contrary, our method would excel at evaluating moves in order not to lose winning strategies because betweenness centrality measures the balance of importance of each cell. To improve our method, it would be better to find different network characteristics for the second player.

A second possible reason is that prior knowledge was implemented in Wolve, and this might be effective only for the second player. For example, Wolve used *mustplay*. Mustplay involves cells that the player must place on pain of losing the match. Prior knowledge, especially mustplay, is useful for the second player but not for the first player because the latter has winning strategies while the former starts at a disadvantage. For these reasons, by implementing prior knowledge, the winning percentage of our method can be further improved.

6   Conclusion

In this paper, we proposed a novel evaluation function for computer Hex that can take into account the global and local strategies calculated from network characteristics on the board network. The improved method can dynamically change the weight of the global and local strategies. We showed that our improved method is superior to Wolve, which uses the currently popular evaluation function based on an electrical circuit model. However, the winning percentage of the second player in our method was worse than that in Wolve. In future research, we intend to study methods to improve the winning percentage of the second player in our Hex strategy, which might require different network characteristics from the ones we considered here, or even prior knowledge. We also intend to consider the swap rule.

References

1. Browne, C.: Hex Strategy: Making the Right Connections. A.K. Peters Ltd, Natick (2000)
2. Even, S., Tarjan, R.E.: A combinatorial problem which is complete in polynomial space. J. Assoc. Comput. Mach. **23**(4), 710–719 (1976)
3. Gale, D.: The game of hex and the brouwer fixed-point theorem. Am. Math. Mon. **75**(10), 818–827 (1979)
4. Henderson, P.: Playing and Solving the Game of Hex, Doctoral Dissertation, pp. 1–149 (2010)
5. Hayward, R.B.: Six wins hex tournament. ICGA J. **29**(3), 163–165 (2006)
6. Anshelevich, V.V.: Hexy wins hex tournament. ICGA J. **23**(3), 181–184 (2000)
7. Huang, S., Arneson, A., Hayward, R.B., Müller, M., Pawlewicz, J.: MoHex2.0: a pattern-based MCTS hex player. Comput. Games, 60–71 (2013)
8. Hayward, R.B.: Mohex wins hex tournament. ICGA J. **35**, 124–127 (2012)
9. Takada, K., Honjo, M., Iizuka, H., Yamamoto, M.: Development of computer hex strategy using network characteristics. Inf. Process. Soc. Jpn **55**(11), 2421–2430 (2014)
10. Hayward, R.B., Arneson, B., Henderson, P.: Mohex wins hex tournament. ICGA J. **32**(2), 114–116 (2009)
11. Hayward, R.B.: Mohex wins hex tournament. ICGA J. **36**(3), 180–183 (2013)
12. Arneson, B., Hayward, R.B., Henderson, P.: Solving hex: beyond humans. In: van den Herik, H.J., Iida, H., Plaat, A. (eds.) CG 2010. LNCS, vol. 6515, pp. 1–10. Springer, Heidelberg (2011)
13. Anshelevich, V.V.: A hierarchical approach to compute hex. Artif. Intell. **134**(1–2), 101–120 (2002)
14. King, D.: Hall of Hexagons (2014). http://www.drking.org.uk/hexagons/hex/index.html
15. Corters, C., Vapnik, V.: Support vector networks. Mach. Learn. **20**, 273–297 (1995)
16. Karatzoglou, A., Smola, A., Hornik, K.: Kenel-Based Machine Learning Lab, R Methods, Version 0.9–20 (2015)
17. Iida, H., Sakuta, M., Rollason, J.: Computer shogi. Artif. Intell. **134**(1–2), 121–144 (2002)

Machine-Learning of Shape Names
for the Game of Go

Kokolo Ikeda[(✉)], Takanari Shishido, and Simon Viennot

Japan Advanced Institute of Science and Technology, Nomi, Japan
{kokolo,sviennot}@jaist.ac.jp

Abstract. Computer Go programs with only a 4-stone handicap have recently defeated professional humans. Now that the strength of Go programs is sufficiently close to that of humans, a new target in artificial intelligence is to develop programs able to provide commentary on Go games. A fundamental difficulty in this development is to learn the terminology of Go, which is often not well defined. An example is the problem of naming shapes such as Atari, Attachment or Hane. In this research, our goal is to allow a program to label relevant moves with an associated shape name. We use machine learning to deduce these names based on local patterns of stones. First, strong amateur players recorded for each game move the associated shape name, using a pre-selected list of 71 terms. Next, these records were used to train a supervised machine learning algorithm. The result is a program able to output the shape name from the local patterns of stones. Including other Go features such as change in liberties improved the performance. Humans agreed on a shape name with a rate of about 82 %. Our algorithm achieved a similar performance, picking the name most preferred by the humans with a rate of about 82 %. This performance is a first step towards a program that is able to communicate with human players in a game review or match.

1 Introduction

Until recently, the research about the game of Go was focused on obtaining strong programs, but the best programs are now able to win against professionals with a 4-stone handicap, which is a level sufficient to play against most amateur players. Apart from strength, a new target for research is now to entertain human players or teach them how to improve at the game. Beginner players often learn the game from strong players, but strong players are not all good teachers, especially because other skills than strength are required for entertaining or teaching. There are not so many skillful teachers and they are expensive, so there is a high need for programs that would be able to teach the game to the players. Ikeda et al. proposed the following 6 requirements for such a program: (1), acquire an opponent model (2), control the advantage of the board position (3), avoid unnatural moves (4), use various strategies (5), use a reasonable amount of time for each move and resign at the correct timing, and (6) comment on the game after it is completed [1].

© Springer International Publishing Switzerland 2015
A. Plaat et al. (Eds.): ACG 2015, LNCS 9525, pp. 247–259, 2015.
DOI: 10.1007/978-3-319-27992-3_22

In this research, we are interested in the last requirement of making the program able to comment on the moves of the player. We consider only a sub-problem, which consists of using the correct shape names (like Attachment or Hane) to refer to game moves. Since the Go board is quite large (usually 19×19), Go players usually refer to moves with such shape names instead of the coordinates that are used in many other games. Handling correctly these shape names is a pre-requisite step for generating comments by a program, but it is also useful in itself to help beginner players understand and memorize the names of the shapes. In this research, we use machine learning to learn the shape name associated to a given move and a given board state. The target game here is the game of Go, but it must be noted that a similar method could be applied to other games, for example to Chess, to output the name of tactical moves such as "fork" or "discovered attack" from the board state.

2   Related Works

With the improvement of algorithms and hardware performance, the level of computer players has now reached a sufficient level of strength for most games, and more research is now done in the field of entertainment and naturalness. For example, in the case of Mario Bros, which is a representative side-scrolling game, there are computer competitions not only for finishing the level as fast as possible, but also for playing as human-like as possible, or for generating levels as fun as possible for humans [2].

Ikeda et al. proposed 6 requirements for entertaining players at the game of Go, with a concrete approach for playing varied strategies or avoiding unnatural moves [1], but no concrete approach was proposed for the requirement of comments or more generally communication with the players.

A trial mode of discussion and comments can be found in some commercial programs. For example, in "Yasashii Igo", the computer players are able to speak and can use to some extent the shape names instead of the coordinates, with sentences like "The cut is a good move". In the recent program "Tencho no Igo 5", the shape of the moves can be read by the voice of a professional female player, which helps to improve the user experience [3]. It is not publicly disclosed how these programs transform a move into the corresponding shape name, but it is probable that they are using a rule-based system with a list of conditional statements. One goal of our research is to develop a reproducible and systematic method for such systems.

If it was possible to associate perfectly a shape name to a move in a given board position, rule-based systems could be sufficient. However, there are many shapes for which the difference is subtle and difficult to express simply, such as "Magari vs Osae", "Nobi vs Hiki", "Tsume vs Extension". A classical way to face such a kind of problem is to use supervised machine-learning [4]. In supervised machine learning, a large set of inputs with the associated correct outputs is needed, and then the parameters of some function model that relates the input to the output are optimized. For example, from a list of 100 examples of Nobi and Hiki, a machine-learning method can possibly find the relation between the board positions and these shape names, and find the right shape names for a given unknown board position, even if the definition of Nobi and Hiki is not perfectly clear.

Machine-learning can be done through many possible algorithms, that can be applied to learn many candidate function models between the input and the output. Some important factors to consider are the number of different elements of the input, and whether the input and the output are discrete or continuous variables. In this research, machine-learning is done with decision trees, but other classical algorithms like neural networks or support vector machines could be considered too. Decision trees have the advantage of giving a result that can be analysed, making it possible to know for what set of conditions on the inputs a given output will be obtained.

3   Proposed Approach

In this research, our goal is to create a program able to tell the shape name of a move, when given a move in a board position. The research is done in the following order, with the same approach that we described in a technical report in Japanese [5].

1. First, we evaluate the possibilities of existing programs to tell the name of moves. TENCHO IGO 5 IS ONE STRONG PROGRAM THAT HAS THE ABILITY OF TELLING THE SHAPE NAME OF THE MOVES, SO WE EVALUATE ITS PERFORMANCE WITH STRONG HUMAN PLAYERS.
2. Next, we perform a supervised machine-learning algorithm for single-move shapes. Some Go terms refer to a set of consecutive moves instead of a single move, but such higher-level terms are not considered in this research and left as a future work. The learning data is a set of shape names each associated to a move and a board position. It is gathered with the help of some strong Go players. Then, we design features that seem promising to distinguish the shapes, like the absolute position or surrounding pattern of stones. We output the value of these features in a file, and the associated shape name given by the human players. Classical machine-learning algorithms can then be performed on this file. We use a simple decision-tree learning algorithm.
3. Then we ask professional players to evaluate the performance of our program at telling the shape names of moves. Shape names are not always unique for a given move, so the professionals were asked to evaluate the shape names in a graded scale of satisfaction, instead of a binary correct/incorrect way.

4   Gathering of Learning Data and Performance of Existing Methods

In this section we describe the limited set of target shapes (Sect. 4.1), the gathering of learned data (Sect. 4.2), learning data (Sect. 4.3), and the performance of existing methods (Sect. 4.4).

4.1   Limited Set of Target Shapes

In this research, we try to learn the names of the most basic and classical
Go shapes. As shown in Table 1, we selected 71 basic shapes to be the target of the learning. For example, upper-Attachment, lower-Attachment, outside-
Attachment, inside-Attachment are not distinguished and are all classified as
an Attachment; high one-space Approach or two-space Approach are not distinguished and are all considered as an Approach. Moreover, terms referring to
the meaning of a move more than the shape are not included in the list, such as
Attack, Defence, Ladder-breaker, Kusuguri, Ko-threat, Yosumi or Kikashi.

4.2   Gathering of Learning Data

The learning data needed for this research consists of a list of board positions,
with the move played in that position and the name of the shape associated to
this move. To gather this learning data, we asked the cooperation of strong Go
players. We asked them to input the name of the shapes with the free software
Multigo [6], which can be used to read and also add comments to the record of
a Go game.

We use an input format such as "Tobi, Extension (90)", that allows to record
not only one name for the shape, but multiple candidates. Multiple candidates
are allowed because it is frequent that the same move can be referred to by
multiple shape names. The number between parenthesis after a candidate name
for the shape is an evaluation between 70 and 100 by the human player of how
much this name is adapted to refer to the targeted shape. The evaluation of
the first candidate is always 100 points and so is not written. The players that
recorded the shape names and their evaluation were told that 90 corresponds to
a name almost as adequate as the first candidate, a value of 80 to a name that
feels not so right, and a value of 70 to a name that feels a bit strange but could
still be possible.

4.3   Learning Data

We describe in this section the details of the gathered learning data. The players
that recorded the shape names are 6 strong players, ranked above 4 dan on
the KGS server. The game records come from 60 games between professionals
or top amateurs players, for a total of 11,526 moves. As shown in Table 1, the
most frequent shape found in these moves is the Connection (Tsugi), found 1404
times, followed by the Osae, 1062 times.

There are big differences between the number of appearances of the shapes in
the game records. Some famous shapes appear less than 10 times, mainly because
despite the fact that they are famous, they appear rarely in a game. It implies
that it will probably be hard to classify them during the machine-learning.

Different game records were dispatched to the game players, except for one
game record (with 117 moves) that was given to all players. The frequency at
which the players recorded the same shape name on this common game record

Table 1. Target shapes and number of appearances

Connection	(1404)	Sagari	(223)	Guzumi	(88)	Kata	(50)	Tsukidashi	(18)
Osae	(1062)	Extension	(209)	Tobitsuke	(87)	Soi	(46)	Wariuchi	(16)
Hane	(940)	Butsukari	(203)	Pincer	(84)	Narabi	(46)	Tobikomi	(10)
Atari	(827)	Hai	(193)	Watari	(80)	Kado	(44)	Keima-connection	(9)
Nobi	(639)	Hiki	(192)	Hanedashi	(67)	Tobisagari	(40)	Counter-pincer	(7)
Push-through	(612)	Ko-tori	(176)	Tori	(67)	Hasamitsuke	(39)	Hoshishita	(7)
Tobi	(575)	Approach	(170)	Shimari	(66)	Ogeima	(37)	Ryokakari	(7)
Cut	(531)	Kaketsugi	(151)	Kake	(66)	Hirakizume	(37)	Hekomi	(6)
Attachment	(441)	Nige	(139)	Uchikomi	(65)	Horikomi	(36)	Geta	(4)
Keima	(386)	Kakae	(135)	Suberi	(64)	Oki	(35)	Takamoku	(4)
Kosumi	(352)	Komoku	(133)	Boshi	(62)	Hazama	(26)	Mokuhazushi	(5)
Nuki	(351)	Fukurami	(123)	Warikomi	(62)	Tachi	(26)		
Oshi	(302)	Hoshi	(105)	Tsume	(60)	Tsukekoshi	(20)		
Nozoki	(295)	Kosumitsuke	(103)	Takefu	(54)	Hanekomi	(18)		
Magari	(251)	Atekomi	(101)	San-san	(54)	Sashikomi	(18)		

was found to be 82.2 % when considering only the first candidate, and only 87.0 % even when considering up to the second candidate. Despite the fact that 5 of the 6 players come from the same community, a difference of almost 20 points of percentage in the name used for a move shows the difficulty of defining and assigning a shape name to a move. Figure 1 shows some examples where the human players did not give the same shape name.

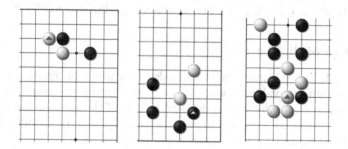

Fig. 1. Shapes labeled with multiple names by human players. Left: Hane, Osae. Middle: Kosumi, San-san. Right: Osae, Push-through, Cut, Magari, Guzumi

4.4   Performance of Existing Methods

In this section, we evaluate the performance of two existing programs for naming the shapes of Go moves. The first program is the popular commercially available TENCHO NO IGO 5, which has a mode for reading the shape names of the moves.

We gave 4 of the game records to the TENCHO program, and it returned the shape name for a total of 262 moves. Then, we asked to a strong Go player to evaluate the names given by TENCHO and to classify them in the following 4 categories: (1) Correct shape name, (2) Unnatural shape name, (3) Absence of shape name, (4) Incorrect shape name.

The 262 moves were divided in these 4 categories as follows. (1) 65.6 %, (2) 2.3 %, (3) 30.2 %, (4) 1.9 %. It shows that the number of mistakes in the shape names is small, but that in many cases, no shape name was given. Especially, high-level shapes such as Atekomi, Guzumi or Hasami-tsuke are not output by the program. We guess that for a commercial software, the priority was given to avoiding mistakes rather than trying to give a shape name to all the moves.

NOMITAN is a Go program developed in Japan Advanced Institute of Science and Technology (JAIST) in the Ikeda and Iida laboratories. The program contains a hand-coded set of 554 conditional rules that outputs the shape name corresponding to a move. When applied on the 11,526 moves of Sect. 4.3, the output of NOMITAN matched the first candidate of human players with a rate of 73.7 %. The rate was 76.6 % when considering up to the second move. Compared to the matching rate of 82.2 % and 87 % between human players, there is almost a drop of 10 percentage points. This feature of Nomitan was originally designed for a 9×9 board [7], and the absence of shapes such as Extension (Hiraki) and Boshi is part of the reason for this drop.

The goal of this paper is to propose an approach that gives better performance than the current existing programs.

5   Machine Learning and Preliminary Experiments

In this section we describe the design of the features (Sect. 5.1), a machine-learning algorithm (Sect. 5.1), and preliminary experimental results (Sect. 5.1).

5.1   Design of the Features

It is preferable for the supervised machine-learning to use as inputs not the raw board state, but the value of some well-chosen abstract features. The chosen features have a direct influence on the performance of the machine-learning. Too rough features prevent a good representation ability, but too detailed features lead to overfitting and a poor generalization performance.

As a starting point, we have used the features already implemented in the rule-based shape naming functions of Nomitan. These features are a simplified version of features initially designed for a strong Go program. The only feature really specific to shape naming is the feature related to the Cut. After removing some useless features, this set of candidate features contains 25 features. The distance used in the features is not the Euclidean distance, but the R-distance defined by $d(\delta x, \delta y) = \delta x + \delta y + \max(\delta x, \delta y)$ as in [8].

- F1 (2 features) (x, y) coordinates so that $y \le x \le 10$ after rotation and reflection. Needed to categorize shapes such as Hoshi and Komoku.
- F2 Line of the stone, i.e., the distance to the closest board edge.

- F3 R-distance to the closest stone of the same color. If there are no other surrounding stones, a distance of 2 means a Narabi, 3 a Kosumi, 4 a Tobi, and 5 a Keima.
- F4 R-distance to the closest enemy stone (opposite color). If there are no other surrounding stones, a distance of 2 is an Attachment, 3 a Kado or Kata.
- F5 Line of the closest stone of the same color.
- F6 Line of the closest enemy stone. Often useful to distinguish Kado and Kata, Oshi and Hai.
- F7 Number of enemy stones with only one liberty left in horizontal or vertical contact. If this value is bigger than zero, the shape is often a Nuki.
- F8 Number of enemy stones with two liberties in horizontal or vertical contact. If this value is bigger than zero, the shape is often Atari.
- F9 Number of stones of the same color with only one liberty left in horizontal or vertical contact.
- F10 Number of stones of the same color with only two liberties left in horizontal or vertical contact.
- F11 Number of liberties of the group of stones containing the stone just played.
- F12 Whether there is a stone of the same color left-down diagonally, and enemy stones on the left and the down. This is directly related to a Cut (Kiri).
- F13 (12 features) State of the 12 board intersections at R-distance 2 to 4 (0: empty, 1: stone of the same color, 2: enemy stone, 3: out of the board).

5.2   Machine-Learning Algorithm

The supervised machine-learning is done with J48 (implementation in Java of C4.5) from the data-mining Weka software [9,10] as follows.

1. The raw gathered data consists of a collection of board positions with their associated moves and corresponding shape names, in the sgf file format. However, Weka cannot use directly such files, so we use first a script in combination with NOMITAN to compute the value of the features described in Sect. 5.1, and output them in a csv file that can be read with Weka.
2. The csv file is read through Weka, and some information useless for the machine learning (such as the game record number or the number of moves) is removed. Then, a decision tree is constructed with J48, and the matching ratio is obtained. It takes less than 1 s on a typical PC, and even the 10-fold cross-validation takes less than 10 s.
3. We also obtain the matching ratio between the decision tree output and the second shape name candidate.

5.3   Preliminary Experimental Results

In order to determine how well our machine-learning method works, we run a preliminary experiment with the features described in Sect. 5.1. The matching ratio after the machine-learning is as follows.

– 75.3 % matching on the first candidate
– 76.8 % matching when considering up to the second candidate

It is already slightly better than the 73.7 % and 76.6 % of the rule-based system of NOMITAN, but still far from the 82.2 % and 87 % of matching ratio between humans.

We compare in Table 2 the results of NOMITAN and of the machine-learning on some shapes related to the surrounding pattern of stones. The machine-learning is clearly not very efficient on these shapes, which shows that some improvement of the features is needed. The shapes appear a few hundred times, so we can expect the machine-learning to work efficiently after a re-design of the features.

Table 2. Proportion of correct answers for shapes related to surrounding patterns

Correctness ratio	NOMITAN	Machine-learning
Magari (251)	76.6	41.6
Push-through (612)	83.2	59.3
Oshi (302)	85.4	65.2

6   Improvement of the Features and Evaluation Experiment

In this section, we describe how we improved the features to solve the problematic shapes presented in the previous section (Sect. 6.1), the relative importance of the features (Sect. 6.2), the proportion of correct shape names and remaining problems (Sect. 6.3), and the evaluation by a professional player (Sect. 6.4).

6.1   Improvement of the Features

The features related to the surrounding pattern of stones use 12 board intersections. The problem of these features is that it does not take into account rotation and reflection equivalences. For example, the 8 patterns of Fig. 2 all correspond to the same 3×3 pattern when rotations and reflections are considered. The corresponding shape name is usually called Push-through (De). If the 8 patterns are considered separately, it increases the number of conditions needed in the decision tree, and more importantly, the quantity of learning data for each pattern is greatly reduced. So, we included rotations and reflections in the design of the features. The 8 patterns are reduced to a canonical pattern with the following order: place under the played stone as many stones from the player as possible, then as many enemy stones as possible; then on the left, the right, the bottom-left, the bottom-right and the top-left.

On the example of Fig. 2, (b) and (e) are given the priority with condition 1 of placing stones of the player below, and then the priority is given to (e) with condition 2 of placing stones of the players on the left. Then, all 8 patterns are represented by the single (e) canonical pattern.

The effect of this improvement was quite important, with an increase of 5 points of the learning matching ratio. The rate of correct answers for the 3 shapes of Table 2 raised considerably to reach Magari: 75.5 %, Push-through: 83.6 %, Oshi: 81.9 %. We also tried to use the Tengen point (central point of the Goban) as a reference in the order, to distinguish patterns oriented differently towards the Tengen, but the improvement of the learning matching ratio was only 1 point in that case. Keeping the information about the orientation towards the Tengen is useful for some shapes such as Oshi and Hai, but for most other shapes, it is more important to group the patterns under a single canonical pattern. It increases the number of learning examples for this canonical pattern.

In addition, we have removed some too-fine features and tuned some parameters. The improvement from each of the following modifications was around 0.3 point so we cannot be sure that all of them would be welcome on a different set of learning data. The six improvements read as follows.

- Feature related to the line of the stone is removed.
- The 8 intersections at an R-distance of 5 are added to the stone pattern.
- After rotation, the 3 intersections above and the 3 intersections below the pattern are added to the pattern.
- After rotation, the number of stones of the player on the bottom-left, left, bottom-right, bottom of the bottom-left, bottom of the bottom, bottom of the bottom-right are added.
- Confidence parameter of J48 is set to 0.1 instead of the default 0.25 value.
- SubTree parameter of J48 is set to false.

After all these improvements, the matching ratio on the first shape name reached 82.0 %, and 85.4 % when considering up to the second shape name. It is significantly better than the performance of the rule-based system of NOMITAN, and quite close to the 82.2 % and 87.0 % matching ratio between humans.

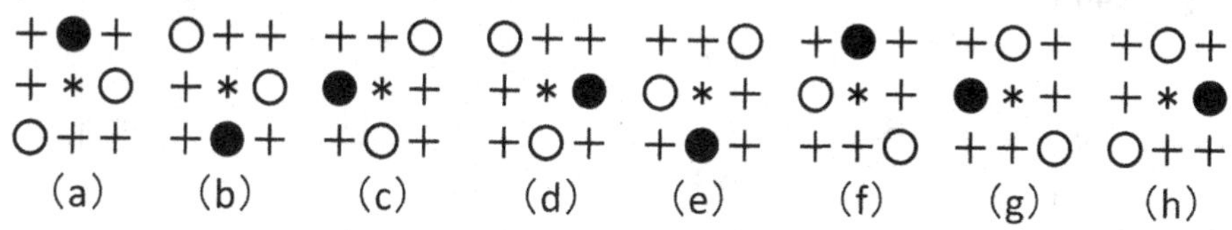

$$
\begin{array}{cccccccc}
+\bullet+ & \bigcirc++ & ++\bigcirc & \bigcirc++ & ++\bigcirc & +\bullet+ & +\bigcirc+ & +\bigcirc+ \\
+*\bigcirc & +*\bigcirc & \bullet*+ & +*\bullet & \bigcirc*+ & \bigcirc*+ & \bullet*+ & +*\bullet \\
\bigcirc++ & +\bullet+ & +\bigcirc+ & +\bigcirc+ & +\bullet+ & ++\bigcirc & ++\bigcirc & \bigcirc++ \\
(a) & (b) & (c) & (d) & (e) & (f) & (g) & (h)
\end{array}
$$

Fig. 2. 8 equivalent patterns, often corresponding to a Push-through (De)

6.2   Relative Importance of the Features

We have also tested the influence on the matching ratio of removing some features or changing the size of the local patterns. If only the local patterns (F13) are used, the matching ratio drops from 82.0 % to 75.3 %. If patterns of size 4 are used instead of 5, it drops to 70.3 %, and then to 61.6 % with patterns of size 3. If only F7 and F8 are removed, the maching ratio drops from 82.0 % to 78.9 %. The recall of Atari and Nuki shapes is significantly decreased by 17 % and 37 %. These shapes are clear examples where local patterns of stones are not sufficient.

6.3   Proportion of Correct Shape Names and Remaining Problems

The global matching ratio of the shape names is around 82 %, but there are big differences between the shapes, some of them being better categorized than others. On Fig. 3, we show the proportion of correct naming for the different shapes, in function of the number of appearances of the shape in the game records (logscale). The general tendency is that the proportion of correct naming increases with the number of examples, but even for a similar number of appearances in the game records, there is a large spreading of the proportion of correct naming between the shapes. For example, for shapes such as Hoshi, Komoku, Takamoku, Mokuhazushi, the proportion of correct naming is almost 100 % even though these shapes appear only a small number of times. The reason is that these shapes are easily identified by their position on the board.

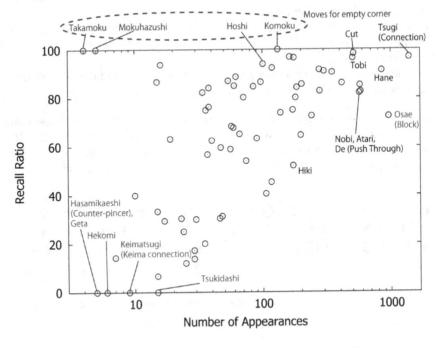

Fig. 3. Proportion of correct names in function of the number of appearances

By contrast, Hirakizume, another shape with a similar low appearance ratio has a very low proportion of correct naming, around 30 %. Figure 4 shows an example named as a Hirakizume by the program, but this shape is clearly not a Hirakizume. The reason for this error is that the rule for Hirakizume in the decision tree is "third or fourth line, with a stone of the same color at R-distance 6 and an enemy stone at R-distance less than 6". These conditions are indeed observed in a Hirakizume shape, but more conditions are needed such as "the stone of the same color and the enemy stone must also be on the third or fourth line". However, there are only 35 appearances of Hirakizume (0.3 %) in the game records, which is probably not sufficient to deduce the full set of conditions characteristic of a Hirakizume shape. This current limitation of the machine-learning could be improved by creating more learning data.

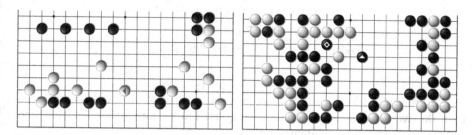

Fig. 4. Examples of naming mistakes. Hirakizume instead of Uchikomi (left), Boshi instead of Nozoki (right)

6.4   Evaluation by a Professional Player

In order to evaluate if the shape naming obtained with the machine-learning is satisfactory from the point of view of humans, we asked to a professional player (Nihon-kiin 6-dan player) to evaluate the shape names recorded in some game records. As explained in Sect. 3, we use a satisfaction scale instead of just a binary correct/incorrect scale.

First, we selected randomly 5 game records from Sect. 4.2 where the shape names were recorded by strong human players. For 3 of these game records, we kept only the first 100 moves, and for the 2 other game records, we kept only the moves from 101 to 200. Then, for this total of 500 moves, we named the shapes with the decision tree obtained from the machine-learning by Weka.

The professional player was not told whether the shape names were recorded by human players or by an algorithm, and was asked to rank the shape name of each move in the following scale.

1. A professional would use the same shape name.
2. A professional would not use this shape name, but it is acceptable.
3. This shape name seems a bit awkward.
4. This shape name is clearly incorrect.

Moreover, the shape name of each move was evaluated with a total score, with roughly the following scale: 90 points if this shape name could be used in a professional review of a game, 80 points if it could be used in a game review by 3-dan amateur players, and 70 points if it could be used by 6-kyu lower-level players. We show the result of the shape name evaluation in Table 3. The columns (2), (3) and (4) show the average number of times in 100 moves where the shape name was evaluated as (2), (3) or (4) by the professional player.

Table 3. Evaluation by a professional of the shape names given by amateur players, and by the machine-learning method. Number of times of bad names and total score.

Game records from	(2)	(3)	(4)	Total score
Human players	5.4	3.8	4.0	84.6
Weka	4.4	3.6	4.6	83.8

The machine learning total score has an average of only 0.8 points lower than the average total score of human players. The performance of the machine learning is close to strong amateur players.

7   Conclusion

In this paper, we presented a method to name automatically the shape of moves for the game of Go. We used machine-learning on learning data recorded by strong human players, who were asked to record the name of each move in a set of game records. Abstract features were used in the machine-learning, and after optimization, the shape naming quality is close to the level of strong amateur players, both in terms of matching ratio and from the point of view of a professional player. There is a limited number of small mistakes, but the main problem is that some big mistakes are found, mainly on shapes that appear rarely. Hopefully, it could be improved by using more learning data, or designing better abstract features. Naming correctly the shape of the moves will probably be important in the future, in order to create programs that are able to teach the game to human players.

Acknowledgment. This work was supported by JSPS KAKENHI Grant Number 26330417.

References

1. Ikeda, K., Viennot, S.: Production of various strategies and position control for Monte-Carlo Go - entertaining human players. In: IEEE-CIG, pp. 145–152 (2013)
2. IEEE-CIG (Computer Intelligence and Games) Competitions. http://geneura.ugr. es/cig2012/competitions.html

3. http://batora1992.blog.fc2.com/blog-entry-17.html
4. Bishop, C.M.: Pattern Recognition and Machine Learning. Springer, New York (2007)
5. Shishido, T., Ikeda, K., Viennot, S.: Japanese expression of the move of Go by machine learning, 33rd GI Kenkyukai, Tokyo (2015)
6. http://www.ruijiang.com/multigo/
7. JAIST CUP 2012, Game Algorithm Competition, 9x9 Entertainment Go Contest. http://www.jaist.ac.jp/jaistcup/2012/jc/9ro.html
8. Coulom, R.: Computing Elo ratings of move patterns in the game of Go. In: ICGA Workshop
9. Quinlan, J.R.: C4.5: Programs for Machine Learning. Morgan Kaufmann Publishers, San Francisco (1993)
10. J48, open source java implementation of C4.5 algorithm. http://weka.sourceforge.net/doc.dev/weka/classifiers/trees/J48.html

Author Index

Printed in the United States
By Bookmasters